Springer Tracts in Modern Physics
Volume 144

W0071777

Springer-Verlag Berlin Heidelberg GmbH

Springer Tracts in Modern Physics

Springer Tracts in Modern Physics provides comprehensive and critical reviews of topics of current interest in physics. The following fields are emphasized: Elementary Particle Physics, Solid-state Physics, Complex Systems, and Fundamental Astrophysics.

Suitable reviews of other fields can also be accepted. The editors encourage prospective authors to correspond with them in advance of submitting an article. For reviews of topics belonging to the above mentioned fields, they should address the responsible editor, otherwise the managing editor.

Managing Editor

Gerhard Höhler

Institut für Theoretische Teilchenphysik
Universität Karlsruhe
Postfach 69 80
D-76128 Karlsruhe, Germany
Phone: +49 (7 21) 6 08 33 75
Fax: +49 (7 21) 37 07 26
Email: gerhard.hoehler@physik.uni-karlsruhe.de
http://www-ttp.physik.uni-karlsruhe.de/
hoehler.html

Elementary Particle Physics, Editors

Johann H. Kühn

Institut für Theoretische Teilchenphysik
Universität Karlsruhe
Postfach 69 80
D-76128 Karlsruhe, Germany
Phone: +49 (7 21) 6 08 33 72
Fax: +49 (7 21) 37 07 26
Email: johann.kuehn@physik.uni-karlsruhe.de
http://www-ttp.physik.uni-karlsruhe.de/~jk

Thomas Müller

Institut für Experimentelle Kernphysik
Fakultät für Physik
Universität Karlsruhe
Postfach 69 80
D-76128 Karlsruhe, Germany
Phone: +49 (7 21) 6 08 35 24
Fax:+49 (7 21) 6 07 26 21
Email: thomas.muller@physik.uni-karlsruhe.de
http://www.ekp.physik.uni-karlsruhe.de

Roberto Peccei

Department of Physics
University of California, Los Angeles
405 Hilgard Avenue
Los Angeles, CA 90024-1547, USA
Phone: +1 310 825 1042
Fax: +1 310 825 9368
Email: peccei@physics.ucla.edu
http://www.physics.ucla.edu/faculty/ladder/
peccei.html

Solid-state Physics, Editor

Peter Wölfle

Institut für Theorie der Kondensierten Materie
Universität Karlsruhe
Postfach 69 80
D-76128 Karlsruhe, Germany
Phone: +49 (7 21) 6 08 35 90
Fax: +49 (7 21) 69 81 50
Email: woelfle@tkm.physik.uni-karlsruhe.de
http://www-tkm.physik.uni-karlsruhe.de

Complex Systems, Editor

Frank Steiner

Abteilung für Theoretische Physik
Universität Ulm
Albert-Einstein-Allee 11
D-89069 Ulm, Germany
Phone: +49 (7 31) 5 02 29 10
Fax: +49 (7 31) 5 02 29 24
Email: steiner@physik.uni-ulm.de
http://www.physik.uni-ulm.de/theo/theophys.html

Fundamental Astrophysis, Editor

Joachim Trümper

Max-Planck-Institut für Extraterrestrische Physik
Postfach 16 03
D-85740 Garching, Germany
Phone: +49 (89) 32 99 35 59
Fax: +49 (89) 32 99 35 69
Email: jtrumper@mpe-garching.mpg.de
http://www.mpe-garching.mpg.de/index.html

POAN Research Group (Ed.)

New Aspects of Electromagnetic and Acoustic Wave Diffusion

With 31 Figures

Springer

POAN Research Group/Groupement de Recherches POAN

Propagation des Ondes en Milieux Aléatoires et/ou Non-Linéaires
and G1180 CNRS
École Polytechnique/ENSTA
Centre de l'Yvette
F-91120 Palaiseau
France

Physics and Astronomy Classification Scheme (PACS):
03.40.Kf, 41.20.Jb, 43.20.Bi, 42.25.Bs/Hz, 42.68.Wt, 43.20.+g, 84.40.Cb, 91.30.Dk

ISSN 0081-3869

ISBN 978-3-662-14757-3   ISBN 978-3-540-69692-6 (eBook)
DOI 10.1007/978-3-540-69692-6

Cataloging-in-Publication Data applied for.

Die Deutsche Bibliothek – CIP-Einheitsaufnahme

New aspects of electromagnetic and acoustic wave diffusion / POAN Research Group (ed.). – Berlin; Heidelberg;
New York; Barcelona; Budapest; Hong Kong; London; Milan; Paris; Santa Clara; Singapore; Tokyo: Springer, 1998
(Springer tracts in modern physics; Vol. 144)

Originally published by Springer-Verlag Berlin Heidelberg New York in 1998.
Softcover reprint of the hardcover 1st edition 1998

Typesetting: Camera-ready copy by the authors using a Springer T$_E$X macro-package
Cover design: *design & production* GmbH, Heidelberg
SPIN: 10669555   56/3144-5 4 3 2 1 0 – Printed on acid-free paper

Preface

Multiple scattering of waves is a rapidly growing field, with many practical applications. At the end of the fifties a dramatic change in attitude occurred towards the physics of multiple scattering of waves. Before that time hardly any attention was given to this phenomenon, because it was believed that multiple scattering washed out the phase of the waves and no interference could occur. Multiple scattering was described by an equation of radiative transfer, equivalent to the Boltzmann equation for classical particles. This equation was successfully used to describe energy transport in stellar atmospheres and neutron propagation in thermal reactors, and there was no need to go beyond. Problems were of a technical and numerical nature, related to the advanced methods necessary to solve the radiative transfer equation accurately [1, 2].

The dramatic switch was initiated by the difficult but widely quoted paper of Anderson in 1958 [3], in which he demonstrated that interference in multiple electron scattering can induce localization effects: the diffusion constant of the wave can vanish. This important conclusion gave an entirely new view on the physics of metal–insulator transitions, since the diffusion constant is closely related to the electronic conductance. After this discovery, the field exploded. The theoretical explanation of superconductivity was given in [4], and other new effects in the conductance were discovered and explained as wave phenomena, such as weak localization [5–7] and the associated Sharvin effect ($h/2e$ oscillations) [8, 9], the Aharanov–Bohm h/e oscillations [10], the quantum Hall effect [11], and, finally, universal conductance fluctuations [12, 13]. These great discoveries led to Nobel prizes for Anderson (1977), Mott (1977), von Klitzing (1985), Bardeen (1956/1972) and Josephson (1973). Today, the domain of mesoscopic wave physics, where multiple scattering and phase feature next to each other, has become one of the most popular topics in physics.

These successes have not gone unnoticed by other physicists. Disordered media have now come into the picture in many branches of physics. In other domains, such as lidar, remote sensing, seismology and medical diagnostics, disorder is an inevitable complication, and recent advances have enabled new progress in the field. A collaboration within a research group facilitates and stimulates the exchange of methods and ideas between several domains of physics. Such cooperation on a national level is the primary aim of the French research group POAN, which realizes that high-quality research is only possible in an international context. We try to realize this aim by organizing interdisciplinary sessions, to which we invite internationally recognized physicists specialized in several disciplines and, preferentially, working on the border between two different disciplines. This work was initiated during the Cargèse conference, held on 21–27 May 1996.

Five principal themes are present in this work: applied mathematics, optics, acoustics, lidar and seismology. The authors have tried to use the same notation and terminology as much as possible. This work should not be considered as a collection of research reports, but rather as a collective effort to give a quick but broad snapshot of the rapidly evolving domain of multiple scattering of waves. The aim is to broaden the discussion on this subject, to arrive at a common language and to facilitate the exchange of ideas between the different communities.

This effort has been made possible by the Groupement de Recherches (GDR) POAN. Scientific contributions have been made by the following people: Professor K. Aki, Dr K. Busch, Professor M. Campillo, Dr Y. Castin, Dr A. Derode, Professor M. Fink, Dr P. Flamant, Dr J.P. Fouque, Professor A.Z. Genack, Professor J. Lacroix, Professor Dr A. Lagendijk, Dr G. Ledanois, Professor Dr G. Maret, Professor R. Maynard, L. Margerin, Dr F. Nicolas, Dr P. Roux, Dr G.L.J.A. Rikken, Dr P. Sebbah, Professor Ping Sheng, Professor C.M. Soukoulis, Dr J.M. Thomas, Dr A. Tip, Dr J.M. Tualle, Dr J. Virmont and Dr D.S. Wiersma. I would like to thank them all for their contributions to this ambitious initiative.

This work was made possible by financial support from the Centre National de Recherche Scientifique (CNRS), Commissariat à l'Énergie Atomique (CEA) and Direction de la Recherche et de la Technologie (Ministère de la Défense) DRET. We thank Annie Touchant, Dr Elisabeth Dubois-Violette, Françoise Berthoud and Michelle Perretto for their valuable help during the preparation of the Physics School in Cargèse, where the ideas leading to this work originated.

Grenoble, March 1998   *Bart van Tiggelen*

Contents

1. Multiple Scattering of Light

Although it has been well recognized for a long time that light and electrons obey more or less the same wave equations, it was not until the beginning of the eighties that interference effects in multiple scattering of light were first considered. Physicists suddenly realized that many advances in solid-state physics applied to light waves as well. During the eighties many physicists were busy translating electronic phenomena to optical phenomena. A good example is the "theory of white paint" by Anderson [14].

In 1985 coherent backscattering of light was experimentally verified by groups in Grenoble [15], Amsterdam [16] and Tel Aviv [17], after preliminary work in the USA [18]. It was the first indication that the equation of radiative transfer is incomplete for electromagnetic waves, since it does not incorporate the phase of the light wave. This opened an entirely new discipline in optics, which is still growing rapidly. Apart from this coherent backscattering effect, the phase of the light is also responsible for a complicated statistical speckle pattern in both transmission and reflection. Detailed experimental and theoretical studies have been carried out on how multiple scattering affects this speckle pattern. The impacts on multiple scattering of gain [19], external magnetic fields [21–23], lower dimensionality [24], nonlinearities [25, 26], Brownian motion [29] (for a recent review see [30]), and dielectric anisotropy such as in liquid crystals [31] have been investigated. Coherent backscattering has also been reported for acoustic waves [32]. This has led to a fundamentally new picture of light diffusion, in which the phase plays a key role. In fact, when the phase becomes part of the physics, diffusion is much more difficult to understand (and to prove), because the association with a classical random walker has disappeared. Despite these novel advances, the holy grail in optics, namely the onset of strong localization of light as predicted by Anderson for electrons, has yet to be verified experimentally, although reports exist for microwaves in quasi-one dimension [33] and two dimensions [34].

The role of symmetry has become an important issue in discussing optical phenomena in the multiple scattering of light. Symmetry acts as a guide when the system becomes more complicated, and gives insight without detailed calculations. Symmetry also enables one to see analogies and differences between different domains. Time-reversal symmetry arguments give rise to interesting comparisons of acoustics and optics. With respect to this aspect acoustics is now ahead of optics. The coherent backscattering peak mentioned above can best be understood on the basis of the reciprocity principle [35]. Such arguments then automatically lead to the prediction that the phenomenon will be destroyed in an external magnetic field [22]. A few years ago, this prediction was verified experimentally in Grenoble by Erbacher, Lenke and Maret [21]. They showed that the interference peak is almost completely destroyed if a magnetic field of 23 T is applied. The enhancement factor of two in coherent backscattering, which seems to be required by the reciprocity principle in the absence of a magnetic field, was recently reconsidered thanks to a very accurate measurement of the coherent backscattering cone carried out by the group in Amsterdam [36]. This measurement also revealed the nonanalytic, triangular shape of the line profile near backscattering, as predicted first by Akkermans et al. [20]. Symmetry arguments like the Onsager relations for transport coefficients allow the existence of a "photonic Hall effect" in the presence of a magnetic field, predicted in 1995 [37], and verified experimentally 1996 in Grenoble [38].

The first application of the phase in multiple scattering of light was developed by Wolf and Maret, who realized that the generalization of quasi-elastic light scattering to multiple-scattering systems enables one to monitor extremely small timescales [29]. The point is that for a light-scattering sequence involving n collisions, the sensitivity of the phase to small movements of the scattering particles is roughly $\sqrt{n}$ times larger. This method, called diffusing-wave spectroscopy, has successfully been applied to measure time correlation functions for all sorts of particle flow (Brownian motion, shear flow and acoustically driven oscillatory motion [39–41]), and is now being applied for imaging purposes in biological tissues [42]. It still awaits applications in atmospheric, colloidal and perhaps even stellar studies.

Many imaging techniques are still based on the conventional inverse problem of the radiative transfer and diffusion equation. Speckle studies go *intrinsically* beyond the radiative transfer equation. The latter describes the average radiation intensity in the medium, averaged over all possible positions of the particles. To describe speckles one has to resort to the autocorrelation function of two intensities at a different frequency, different time, different angle or different position. Another way is to address the whole distribution $P(I)$ of the intensity. The radiative transport equation is nothing more than a low-disorder approximation of only the first moment of the distribution. The common belief is more and more that phase and speckle studies provide a lot of information about what is happening inside a diffuse medium. Genack et al. have measured the phase of multiply scattered microwaves in a direct way [43]. Heterodyne detection methods, used by Jarry et al. [44] and Kempe et al. [45], enable one to monitor the phase at optical frequencies in the multiple-scattering regime.

1.1 Analogies between Light and Electrons?

Bart van Tiggelen and Ad Lagendijk

The analogies between light and electrons are made most obvious by ignoring the spin of the electron and the polarization of light. The wave equations then read

$$i\hbar\partial_t\psi(\mathbf{r},t) = -\frac{\hbar^2}{2m_e}\nabla^2\psi(\mathbf{r},t) + V(\mathbf{r})\psi(\mathbf{r},t) \quad \text{(electrons)}, \tag{1.1}$$

$$-\frac{\varepsilon(\mathbf{r})}{c_0^2}\partial_t^2\psi(\mathbf{r},t) = -\nabla^2\psi(\mathbf{r},t) \quad \text{(light)}. \tag{1.2}$$

In these equations, $V(\mathbf{r})$ is the electron potential and $\varepsilon(\mathbf{r})$ the constitutive dielectric constant of the medium. In Table 1.1 we summarize the analogies between both wave equations.

From Table 1.1 one gets the impression that light and electrons are indeed rather "analogous", implying that results obtained for either one of them seem to apply to both. Only the charge seems to be a fundamental difference, so that any effect intrinsically related to charge will not be automatically "analogous". Closer inspection of Table 1.1 reveals a second difference. Unlike the current density, the density ("conserved quantity") seems rather different. In a random medium (where the position of the particles having either an electronic potential or a dielectric constant are given by a random variable, for instance a Poisson distribution), both the field amplitude $\psi(\mathbf{r},t)$ and the dielectric constant are random variables, and are in fact strongly correlated. This implies that the average densities $\langle\varepsilon\rangle\langle|\psi|^2\rangle$ and $\langle\varepsilon|\psi|^2\rangle$ are not the same.

Table 1.1. Comparison of the equations of motion for Schrödinger potential scattering (modeling electrons without spin) and for scalar dielectric scattering (describing light without polarization, or acoustic waves). For clarity, several fundamental constants have been normalized

Electrons without spin		Light without polarization
$\psi(\mathbf{r},t)$	Wave function	$\psi(\mathbf{r},t)$
$[p^2 + V(\mathbf{r})]\psi = i\partial_t\psi$	Equation of motion	$[p^2 + \varepsilon(\mathbf{r})\partial_t^2]\psi = 0$
$[p^2 + V(\mathbf{r})]\psi_E = E\psi_E$	Eigenvalue equation	$[p^2 + V_E(\mathbf{r})]\psi_E = E\psi_E$
$V(\mathbf{r})$	Potential	$V_E(\mathbf{r}) = [1 - \varepsilon(\mathbf{r})]E$
E: energy	Eigenvalue	$E = \omega^2$: square of frequency
$|\psi|^2$: probability density	Conserved quantity	$\frac{1}{2}\varepsilon(\mathbf{r})|\partial_t\psi|^2 + \frac{1}{2}|\partial_\mathbf{r}\psi|^2$: energy density
$\mathrm{Im}\,\psi^*\partial_\mathbf{r}\psi$	Current density	$-\mathrm{Re}\,\partial_t\psi^*\partial_\mathbf{r}\psi$
$G(z,p) = (z - p^2 - V)^{-1}$	Amplitude Green's function	$G(z,p) = (\varepsilon z^2 - p^2)^{-1}$
Yes	Charge?	No

The difference between both can be shown to be mathematically equivalent to the "energy dependence" of the "optical" potential V_E, which can be identified as the third striking difference between the two kinds of wave [46, 47].

When is this fundamental difference revealed? To answer this question, two approaches can be followed [47]. First, we recall the equation of continuity,

$$\partial_t \text{ density} + \nabla \cdot \textbf{current} = \text{source}, \tag{1.3}$$

to infer that the density is a physical quantity that emerges only in a dynamic experiment, where $\partial_t \neq 0$. Alternatively, one can say that an energy dependence of the potential is only relevant if more than one energy level is involved, by definition a nonstationary situation. Both approaches lead to the same conclusion that light and electrons are *not* analogous in dynamical situations.

In homogeneous media this statement is just a difficult way of saying that the dispersion relation is linear for light ($\omega(\textbf{k}) = kc_0/\sqrt{\varepsilon}$) and parabolic for electrons ($E(\textbf{k}) = \hbar^2 k^2/2m_e$). But what are the consequences of this statement in inhomogeneous media, where the waves are multiply scattered? To this end, let us consider the ensemble-averaged light intensity. On large time and length scales, this intensity is theoretically and experimentally known to obey a simple diffusion equation. The diffusion coefficient is the crucial link between macroscopic and microscopic physics, and is given by the classical formula

$$D(\omega) = \frac{1}{3} v_E \ell^*, \tag{1.4}$$

that is, basically a "transport velocity" v_E and a transport mean free path ℓ^*. This equation looks deceptively simple and familiar in the sense that it contains complicated wave mechanics in a rather familiar way. The transport velocity seems to characterize the dynamics of the diffusion process, and should therefore be different for light and electrons. Despite the rather straightforward argument that leads to this conclusion, the exact expression for the transport velocity of classical waves has only recently been found by the Amsterdam group, after a persistent discrepancy between stationary and dynamic experiments [48]:

$$v_E = \frac{c_0^2}{v_p} \left(1 + n \int_S d\textbf{r} \left[\varepsilon(\textbf{r}) - 1 \right] |\psi_\omega(\textbf{r})|^2 \right)^{-1}. \tag{1.5}$$

In this equation n is the (average) number density of the scatterers (assumed all to be the same), S is the volume of one scatterer and $\psi_\omega(\textbf{r})$ is the eigenfunction of the wave equation (1.2) at frequency ω.

Before this equation was published, people always used the phase velocity v_p – obtained from elementary or more rigorous effective-medium estimates – as the transport velocity, on the basis of a never-justified analogy argument

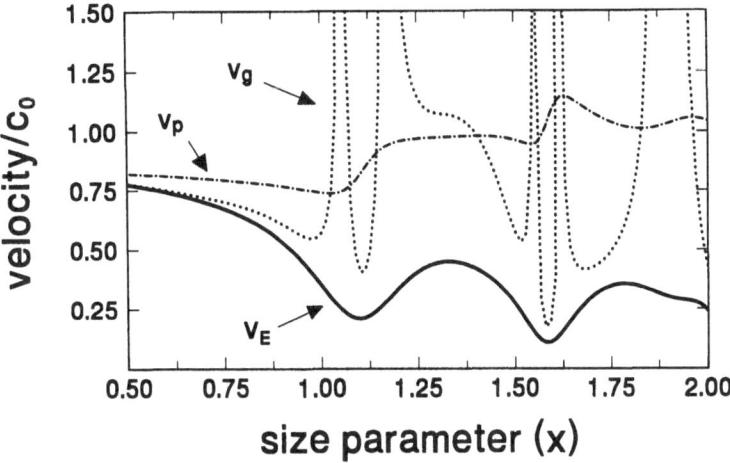

size parameter (x)

Fig. 1.1. Transport velocity, group velocity and phase velocity for a medium with Mie scatterers with index of refraction $m = 2.73$ (corresponding to TiO_2) and size parameter $x = ka$ [reprinted from Lagendijk and van Tiggelen, Physics Reports **270**, 143 (1996), with kind permission from Elsevier Science NL, Saraburghartstraat 25, 1055 KV Amsterdam, The Netherlands]

between light and electrons. The correction term in (1.5) is identified as the stored electromagnetic energy inside a scatterer. This energy can be very large near resonant scattering. For this reason, the transport velocity can become very small, as had been indicated first by experiment. In Fig. 1.1 we show a calculation of this velocity, and compare it to two other velocities: the phase and the group velocity. For electron–impurity scattering in the solid state, the correction factor is absent.

After some controversy [49, 50], the nontrivial correction term in the denominator of (1.5) has been confirmed by different calculations by Kogan and Kaveh [51], and Cwilich and Fu [52]. Also, acoustic waves are affected by this correction factor [53]. The transport velocity has been investigated for light in 2D by Busch and Soukoulis [54], who also devised a successful, novel, effective-medium method based on this approach [55], going beyond the low-density approximation. Kroha, Soukoulis and Wölfle [56], as well as Livdan and Lisyansky [57], discuss the transport velocity in a sophisticated transport theory with near-field effects. Consistency between the various approaches has been considered in [46]. Barabanenkov et al. [58] discuss the modification of (1.5) in the more general regime of radiative transfer, where the delay effects can be discussed for all orders of scattering and with a proper account of polarization. Experiments [59] can also monitor how the speed goes from the group velocity in the regime of the coherent beam to the transport velocity in the asymptotic diffuse regime. The transport velocity has also found an application in phonon transport in strongly scattering media [60].

Recently, the theory has been confirmed experimentally for microwaves in dilute samples [61].

1.2 Coherent Beam, Diffuse Beam and Speckles: the Old View

Bart van Tiggelen and Ad Lagendijk

The "old" view of multiple scattering of light is roughly 100 years old. In this section we briefly discuss its essence. We shall discriminate between the different moments of the field amplitude.

1.2.1 Coherent Beam

The coherent beam is mathematically defined as the average field amplitude, and denoted by $\langle \psi(\mathbf{r}, t) \rangle$. By the "average" is meant a statistical average over all possible realizations of the scatterers. Under reasonable assumptions it can be shown that the coherent beam obeys the same wave equation, but with a homogeneous, complex-valued "effective-medium" dielectric constant [62]. The *existence* of such a constant (in three dimensions) can be proven in a mathematically rigorous way [63] and is discussed in Sect. 6.2. The fact that the scattered fields of all the scatterers can be replaced by one field propagating in an effective homogeneous dielectric medium is far from trivial. In optics books it is referred to as the Oseen–Ewald extinction theorem [64]. What is often not mentioned explicitly is the fact that it is a theory for the coherent beam only.

The name "coherent beam" has given rise to a lot of confusion. Owing to momentum conservation, the average field inside a thick slab has more or less the same wave number as an incident plane wave. For this reason, it corresponds physically to the remnant of the incident wave. In fact, the many multiple scattering events in the medium add up to the "coherent" beam, which does not seem to be "scattered" at all. It only decays in space or time, and its phase velocity is changed. The decay in space and time, mathematically due to the averaging, can be associated physically with the randomization of the phase. This explains the word "coherent". In the new view of multiple scattering, the "diffuse beam" (the ensemble-averaged intensity) is recognized to be highly coherent, but for historic reasons one has never changed the name. The coherent beam is often erroneously referred to as "ballistic beam". The latter refers to the beam that is not scattered *at all*. As a result this beam must *necessarily* propagate with the vacuum speed of light.

The complex value of the effective dielectric constant is responsible for an (exponential) decay of the intensity. One expects the "coherent" contribution

$\langle T_{\mathrm{coh}} \rangle$ to the transmission of a medium with plane parallel geometry thickness L, to be

$$\langle T_{\mathrm{coh}} \rangle \sim \mathrm{e}^{-L/\ell}, \tag{1.6}$$

where ℓ is the scattering mean free path of the medium. This formula is known as the Lambert–Beer formula. If the density n of the scatterers is sufficiently low, the familiar result $\ell = 1/n\sigma$ can be obtained, where σ is the total extinction cross-section, including scattering and absorption [62]. The wavelength of the coherent beam also changes in the medium, and is essentially given by the real part of the complex dielectric constant.

The first microscopic expression for this complex constant was obtained by Lorentz [65], one century ago, in addressing the question of a random collection of "pointlike" atoms with frequency-dependent polarizability $\alpha(\omega)$ and average number density n:

$$\varepsilon(\omega) = 1 + \frac{n\alpha(\omega)}{1 - n\alpha(\omega)/3} . \tag{1.7}$$

This equation applies very well for dilute atomic gases. It was later given a more microscopic basis, both at zero frequency [66] and at optical frequencies [67]. The factor $1/(1 - n\alpha/3)$ is often referred to as the local field factor, and stems from the longitudinal fields in the Maxwell equations. For random collections of Mie spheres, (1.7) does not apply and sophisticated calculations [68, 69] have been carried out.

Some confusion existed as to what velocity the coherent beam actually propagates with. A dynamic analysis for a wave packet that nearly resembles a plane wave yields the group velocity $v_{\mathrm{g}} = \mathrm{d}\omega/\mathrm{d}k$ [64]. In the presence of strong resonant scattering, this group velocity becomes negative or even complex-valued, and seems no longer to be a physical velocity. Nevertheless, theory [70] and experiment [71] with optical waves have indicated that the group velocity is a physical velocity, even in the regime of anomalous dispersion. This notion is confirmed for acoustic waves [229], as explained in Sect. 4.2.

1.2.2 Diffuse Beam and Speckles

The phenomenological equation of radiative transfer,

$$\frac{1}{v_{\mathrm{p}}} \partial_t I_{\mathbf{k}}(\mathbf{r}, t) + \hat{\mathbf{k}} \cdot \nabla I_{\mathbf{k}}(\mathbf{r}, t) + \frac{1}{\ell} I_{\mathbf{k}}(\mathbf{r}, t)$$

$$= \mathrm{source} + n \int \mathrm{d}^2 \Omega_{\mathbf{k}'} \frac{\mathrm{d}\sigma}{\mathrm{d}\Omega}(\hat{\mathbf{k}}' \to \hat{\mathbf{k}}) \, I_{\mathbf{k}'}(\mathbf{r}, t), \tag{1.8}$$

describes the average *specific* radiation intensity $I_{\mathbf{k}}(\mathbf{r}, t)$ for light propagating in direction $\mathbf{k}$, at time t and position $\mathbf{r}$ (we drop the averaging brackets if no confusion can arise). The transfer equation can be obtained from the underlying wave equation provided some low-density approximation is made

[72, 47]. Until 20 years ago no substantial disagreement was found with optical experiments.

Of course, physicists knew that speckle spots exist, even in multiple scattering. You can see them in the experiment, even with the naked eye, but they are not given by (1.8). To describe them seems to pose no problem. The intensity in some scattering channel (speckle spot) a is basically $I(a) = \psi(a)\psi^*(a)$, and by noting that in multiple scattering the field $\psi(a)$ is in fact a complex random variable given by a sum of many partial and decorrelated waves, elementary arguments demonstrate that the probability distribution of the radiation intensity is a Poisson distribution [73], e.g.

$$P(I) = \frac{1}{\langle I \rangle} \exp\left(-\frac{I}{\langle I \rangle}\right). \tag{1.9}$$

This expresses all moments in terms of the average intensity, found by solving (1.8). In the simplest formulation, a speckle can be quantified by its first cumulant, for which (1.9) gives $\Delta I = \langle I \rangle$. The same random-phase arguments can be repeated for the autocorrelation function of the intensity in speckle spots at different frequencies, different angles, different positions or different times. The result is

$$\langle I_a I_b \rangle = \langle I_a \rangle \langle I_b \rangle + |\langle \psi_a \psi_b^* \rangle|^2, \tag{1.10}$$

known as the Siegert relation [74]. To calculate the correlator $\langle \psi_a \psi_b^* \rangle$, only variants of the radiative transfer equation are needed. Again, until some 10 years ago, this approach was more than sufficient to describe the experiments.

These days, very sophisticated and efficient numerical methods exist to solve the equation of radiative transfer, even with polarization [2, 75]. However, in view of the complexity one often prefers to make the so-called diffusion approximation, in which one replaces (1.8) by a diffusion equation, assuming that the specific radiation intensity can be expressed in terms of only the local current density $\mathbf{J}$ and the local radiation density Φ. For an isotropic medium this approximation reads [76]

$$I_\mathbf{k}(\mathbf{r}, t) \sim \Phi(\mathbf{r}, t)c + 3\hat{\mathbf{k}} \cdot \mathbf{J}(\mathbf{r}, t). \tag{1.11}$$

With this approximation the relation between current density and density becomes, upon taking the first angular moment of the radiative transfer equation,

$$\mathbf{J} = -D\nabla\Phi, \tag{1.12}$$

with D given by (1.4). In particular, the transport mean free path can be shown to be equal to

$$\ell^* = \frac{\ell}{1 - \langle \cos\theta \rangle}. \tag{1.13}$$

Note the subtle difference between the transport mean free path obtained for the average intensity and the scattering mean free path, which determines the decay of the coherent beam. Unlike to the scattering mean free

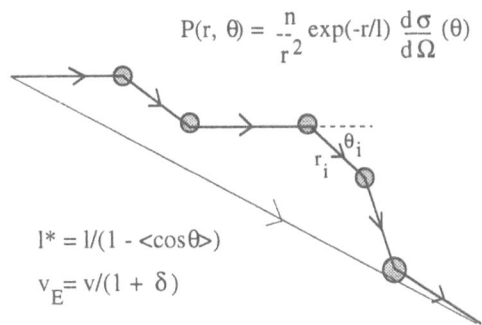

$$P(r, \theta) = \frac{n}{r^2} \exp(-r/l) \frac{d\sigma}{d\Omega} (\theta)$$

$$l^* = l/(1 - \langle \cos\theta \rangle)$$

$$v_E = v/(1 + \delta)$$

$$\lim_{t \to \infty} \frac{(\Sigma_i x_i)^2}{t} = 2\, d\, D = 2\, v_E l^*$$

Fig. 1.2. Stochastic interpretation of the diffusion picture. In this picture, the probability for a photon to travel – after a scattering event – a distance r in a direction θ is given by the normalized distribution $P(r, \theta)$, being the product of the scatterer density n, an exponential and the differential cross-section of one scatterer. The exponential step-length distribution defines the scattering mean free path in the random medium. The "Brownian-motion" definition of the diffusion constant in d dimensions is $D = r^2/2dt$ for large t, and given by the product of a transport velocity and a transport mean free path. Both are affected by the scattering. As indicated, the transport mean free path ℓ^* becomes longer than the scattering mean free path if the scattering is predominantly in the forward direction. For the exponential step-length distribution the result $\ell^* = \ell/(1 - \langle \cos\theta \rangle)$ is obtained, which differs from the familiar relation for polymers with fixed "step length" [27, 28]. The transport velocity v_E suffers from scattering delay. The factor δ can be expressed as an angular average of the delay time of waves due to the scattering [48]. It is important to note that this picture is a *probabilistic interpretation* of a transport equation and not the real world: the displayed scatterers are – on average – separated by one mean free path $1/n\sigma$ and not by $n^{-1/3}$

path, the transport mean free path does not count forward scattering as real scattering (Fig. 1.2). From the continuity equation (1.3) one can now derive the diffusion equation for Φ. The stationary solution of this equation for a plane parallel slab with thickness L, accompanied by appropriate boundary conditions, enables us to calculate the diffuse angular transmission [77],

$$\langle T_{\mathrm{incoh}}(L, \theta) \rangle = \frac{2}{3} \frac{\ell^*}{L + \frac{4}{3}\ell^*} \left(1 + \frac{3}{2}\cos\theta\right). \tag{1.14}$$

Note that this decays algebraically as a function of the slab thickness, quite contrary to the almost-coherent component derived in (1.6). It is also interesting to mention that (1.14) can be derived *exactly* from the radiative transfer

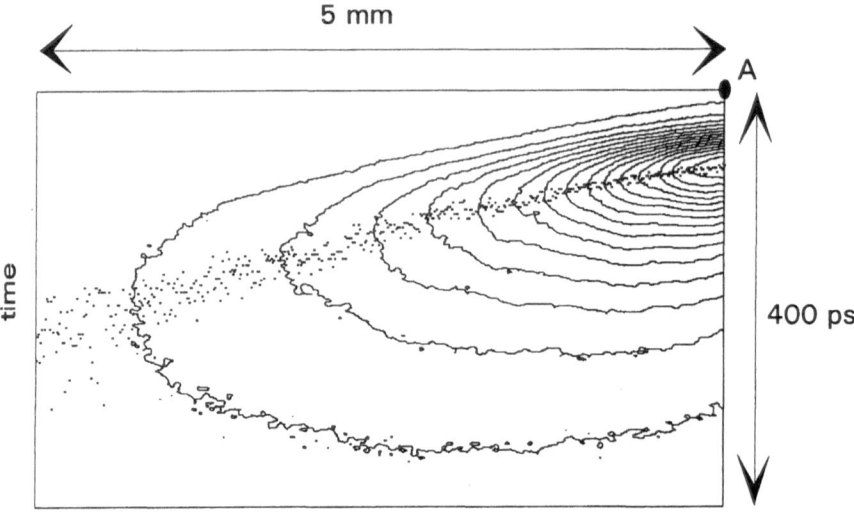

Fig. 1.3. Time- and space-resolved reflectance recorded from a suspension of 435 nm latex microspheres with a 0.97 % volume fraction. The results are shown by means of equi-intensity contours on an arbitrary linear scale. Point A is 2 mm away from the source. The time origin was determined by subsequent analysis of the results. The maxima of reflectance at a given distance from the source are indicated on the figure by dots [reprinted from Tualle, Gelebart, Tinet, Avrillier and Ollivier, Opt. Comm. **124**, 216 (1996), with kind permission from Elsevier Science NL, Saraburghartstraat 25, 1055 KV Amsterdam, The Netherlands]

equation, without resorting to the diffusion approximation at all, where the only difference is that the numerical factor 4/3 in the denominator is replaced by $2z_0 = 1.42 \cdots$. This confirms that the diffusion domain is a genuine asymptotic limit in radiative transfer. Of course, it has some deficiencies, such as the occurrence of negative incident specific intensities near the borders that do not show up in the rigorous transfer equation. Note also that the velocity of the light is absent in this transmission coefficient, as it should be in any stationary physical quantity. We emphasize that in this "old picture" of radiative transfer, the velocity of light is some effective-medium velocity v_p, as evident from (1.8), and not the transport velocity v_E considered in Sect. 1.1.

The validity of the diffusion approximation has been the topic of several studies. Recently, accurate measurements with extremely short pulses and sophisticated detection methods (streak camera) have verified its validity. In Fig. 1.3 we show a measurement by Tualle et al. [78]. Such a measurement enables one to determine the mean free path and the diffusion tensor with great accuracy. In anisotropic media, the diffusion approximation expressed by (1.11) needs to be modified [79–84].

1.3 Diffusing-Wave Spectroscopy

Georg Maret and Roger Maynard

This section focuses on recent work on *temporal* fluctuations of the intensity of multiply scattered light which are caused by motion of the scatterers. While parts of the underlying physics had already been discussed in 1983/84 [85, 86], dynamic multiple scattering of light was introduced by the work on calibrated colloidal latex particles in aqueous suspensions [29, 30, 87], and has since rapidly evolved into a powerful technique called "diffusing-wave spectroscopy"(DWS).

1.3.1 Basic Physics

The principle and mathematical treatment of DWS can be found in various reviews (e.g. [74, 88]). We therefore just briefly summarize the physics. Coherent light waves (of, say, an incident monomode laser beam) travel inside the sample along various random scattering paths described by a photon random walk, and set up at the detector a highly irregular intensity pattern called "speckle" as a result of interference between many waves from many paths of various lengths. As in conventional dynamic single scattering, the intensity in a given speckle spot fluctuates when the scatterers move with respect to each other. Since the transit time of photons along a typical multiple-scattering path is much shorter than the time τ_0 it takes a colloidal particle to move a distance of order of the optical wavelength $\lambda_0 = 2\pi/k_0$ ($\tau_0 = 1/Dk_0^2$ for Brownian motion with diffusion coefficient D), the problem is treated in a quasi-stationary approximation. The time-dependent phase shifts $\varphi(t)$ of the scattered optical fields due to motion of the scatterers accumulate along the paths, giving rise to speckle fluctuations on a path-length-dependent timescale. Consequently, under conditions of strong multiple scattering, this timescale is much faster than τ_0. Unlike single scattering, the timescale does not depend on the angle of observation, but rather on the geometry of the scattering cell, which controls the typical path length and its distribution. The seemingly complicated calculation of measurable quantities such as the frequency spectrum or the time autocorrelation function of the scattered intensity becomes, in fact, rather straightforward in the photon diffusion picture. This is seen in the important autocorrelation function of the scattered field [29], $G_1(\mathbf{r}, t) = \langle E(\mathbf{r}, t_0) E^*(\mathbf{r}, t_0 + t) \rangle$, which can be put into a normalized form $g_1(t)$:

$$g_1(t) = \int\limits_{\ell^*}^{\infty} P(s)\, e^{-(s/\ell^*)\langle \delta\varphi^2(t) \rangle} ds \,\Big/ \int\limits_{\ell^*}^{\infty} P(s)\, ds\,, \qquad (1.15)$$

where $\langle \delta\varphi^2(t) \rangle$ is the mean square phase shift per scattering event and $P(s)$ is a quantity – the path-length distribution – depending on sample geometry, size and transport mean free path ℓ^*, describing how much light intensity is scattered on average into paths of length s. For independent Brownian motion of the scatterers with mean square displacement $\langle \delta r^2(t) \rangle$, we infer that $\langle \delta \varphi^2(t) \rangle = k_o^2 \langle \delta r^2(t) \rangle \approx D k_o^2 t$. Explicit formulas for $P(s)$ and hence $g_1(t)$ have been worked out for various geometries, such as backscattering and transmission from slabs, pairs of optical fibers dipping into a turbid sample, and others, and (1.15) has successfully been tested experimentally on well-characterized colloidal suspensions (e.g. [74, 88]).

Another useful description of the correlation function $G_1(t)$ is related to the solution of the steady-state diffusion equation [89]. In the case of negligible absorption, this equation can be written as

$$\left[-\nabla^2 + k^2(t) \right] G_1(\mathbf{r}, t) = \frac{S(\mathbf{r})}{D_p}, \tag{1.16}$$

where $k^2(t)$ describes the attenuation of temporal fluctuations with time, $D_p = v_E \ell^*/3$ is the photon diffusion constant and $S(\mathbf{r})$ is the light-source distribution. In the case of pure Brownian motion, as described previously, $k^2(t) = 3t/(2\tau_o \ell^{*2})$. This type of analysis can be generalized to the situation of a Poiseuille flow of scatterers, which is simply changing the t dependence of $k^2(t)$.

1.3.2 Specificity of Diffusing-Wave Spectroscopy

DWS has tremendously stimulated the use of light scattering in many fields, in particular in the physics and chemistry of colloids and other complex fluids. *First*, it provides – without the need for index matching – quantitative information about particle displacements $\langle \delta r^2(t) \rangle$ up to concentrations well into the regime of high-order multiple scattering. It works best when single and low-order scattering are negligible and therefore ideally complements other recent techniques such as two-color cross-correlation spectroscopy [90] and single-mode fiber-optic dynamic light scattering [91–93], which essentially suppress the multiple scattered light but still require measurable amounts of single-scattering intensity. DWS is therefore well suited to study interparticle correlations in colloidal suspensions at very high volume fractions and the dynamics of densely packed systems such as concentrated emulsions, foams etc. The *second* important feature of DWS is its extraordinary sensitivity to small displacements of scatterers. In contrast to single scattering of light, which probes fluctuations on length scales larger than $\lambda/2$, displacements as small as $\lambda/1000$ (or even less, in principle) can be monitored with DWS. The probed length scale is easily controlled experimentally by means of the typical path length, i.e. the maximum of $P(s)$, which is set by the sample size and shape and the distance between the injection and detection points of the light. The examples below highlight fluctuations measurable at length scales down to about 0.1 nm, which puts DWS in competition with X-ray and neutron scattering, but covering timescales from tens of nanoseconds to hundreds of seconds. *Third*, DWS experiments on other types of motion, such as shear or oscillatory flow, demonstrate the possibility to characterize flow fields and measure velocity gradients over the experimentally adjustable length scale ℓ^*. *Fourth*, because of the high sensitivity to motion of the scatterers, DWS can detect very small numbers of particles undergoing motion with respect to their surroundings, making it possible to image or localize them even well below the surface of the sample, and to detect sporadic, rare dynamic events. *Last* but not least, DWS is easier to implement experimentally than dynamic single scattering of light because of the intrinsically high scattered intensities and the rather weak sensitivity to misalignment and definition of scattering angle, beam size and polarization.

DWS experiments [94–96] on the short-time crossover from ballistic to Brownian motion of colloidal spheres have clearly revealed a long-time tail in the velocity autocorrelation function and a scaling of its characteristic timescale with the high-frequency shear viscosity of the solution up to high volume fractions, due to hydrodynamic interactions.

DWS is sensitive to relative motions of scatterers other than Brownian motion, as first illustrated by $g_1(t)$ measurements on latex suspensions under Poiseuille flow [97]. If the particle's displacements $\delta \mathbf{r}_i$ are completely correlated because of a deterministic motion as in convective flow, the relevant phase shift $\delta\varphi$ due to two successive scattering events (i) and $(i+1)$ in the expression (1.15) for $g_1(t)$ is $\mathbf{k}_i \cdot (\delta\mathbf{r}_i - \delta\mathbf{r}_{i+1})$. Since $\delta\mathbf{r}_i = \mathbf{v}_i \times t$, it immediately follows that a homogeneous velocity field $\mathbf{v}_i = \text{const.}$ does not generate any temporal speckle fluctuations. Inhomogeneous velocities, however, cause phase fluctuations, thereby generating a decay of $g_1(t)$. The phase fluctuations are given by the velocity difference over the length ℓ^*, since consecutive random scattering events have on average a separation ℓ^*. For homogeneous shear at a rate Γ one again finds the familiar expression (1.15) for $g_1(t)$, the mean square phase shift per scattering event now being $\langle\delta\varphi^2\rangle \approx (\Gamma\ell^* k_0 t)^2$. The t^2 dependence of $\langle\delta\varphi^2\rangle$ – as opposed to the linear t dependence for Brownian motion – is the signature of the deterministic nature of the shear motion. For inhomogeneous shear gradients, such as in Poiseuille flow or plug flow, the decay of $g_1(t)$ becomes somewhat different since the cloud of diffusing photons does not scan the different regions of the flow field with equal weight [98]. Experiments comparing planar flow, Poiseuille flow and Couette flow [99] clearly demonstrate this and are in quantitative agreement with theory. It is thus possible to distinguish between different types of flow and to determine shear gradients in totally turbid liquids by dynamic multiple scattering of light. The Couette flow experiments have been extended to higher shear gradients well into the regime of hydrodynamic instabilities [99]. Beyond a critical shear rate, a characteristic convective roll pattern ("Taylor rolls") appears. The associated additional shear is clearly seen in $g_1(t)$, and scanning the position of a tightly focused incident beam allows one to visualize the otherwise invisible rolls through the position dependence of the characteristic relaxation rate $\Gamma \ell^* k_o$. These experiments are readily extended to turbulent flow, opening the possibility of scale-dependent measurements of $\langle \Gamma^2 \rangle$ [100].

Small longitudinal relative displacements of the particles can also be detected with the help of DWS. This is illustrated by DWS measurements of the variance of the AC electrophoretic mobility in electrorheological fluids [101], and of ultrasound-generated sinusoidal modulation of particle positions, from which the ultrasound amplitude could be estimated optically in solid or liquid multiple-scattering media [102].

1.3.3 Foams and Liquid Crystals

Foams belong to a class of materials with structural features and optical properties very different from those of dense colloidal suspensions, despite their overall "white" appearance in many cases. Dense collections of (air) bubbles are separated by more or less organized soap films and hence the multiple scattering of light cannot be described by scattering from large spheres, but rather should be modeled by multiple reflections from more or less random surfaces. The coarsening and aging of foams have been studied experimentally since 1991 [103, 104], describing the overall slow dynamics as a stochastic sequence of bubble rearrangement events, which are easily detected by the large extension of the diffuse photon cloud, despite the rare occurrence of rearrangements.

Single scattering of light from macroscopically oriented nematic liquid crystals is well understood and is treated in many textbooks. It arises from collective orientation fluctuations of molecules with anisotropic optical polarizability $\delta\epsilon/\epsilon \neq 0$. The statics and dynamics of these fluctuations are described in a continuum-elastic model involving several elastic constants (K) and viscosities (η). In the one-elastic-constant approximation, the amplitude of the light scattered at wave vector q is proportional to $(\delta\epsilon/\epsilon)^2 k_o^4 kT/(Kq^2)$, where kT has the usual meaning. The corresponding relaxation time is $(Kq^2/\eta)^{-1}$, very similar to that of Brownian motion, $(Dq^2)^{-1}$, with a "rotational diffusion constant" $K/\eta \equiv D$. Both scattering amplitude and relaxation time diverge for long wavelengths ($q \to 0$), as it does not cost elastic energy to perform a rotation at $q = 0$. In practice this divergence is avoided by a large-scale cutoff given by the sample size or by a finite electric or magnetic field.

Samples of macroscopically oriented nematic liquid crystals look turbid, although much less so than non-oriented samples. The multiple scattering from unoriented samples has not been studied in detail so far, and thus we focus on dynamic multiple scattering of light studied in [79–84]. In the above model, the low-q divergence of the static structure factor results in a *vanishing* scattering mean free path ℓ, while the transport mean free path ℓ^* stays finite. The photon diffusion constant becomes anisotropic and the orientationally averaged value of ℓ^* is of order $(\delta\epsilon/\epsilon)^2 \, K/(kTk_o^2)$ [79–84]. On scales beyond ℓ^* we therefore recover an anisotropic photon diffusion picture (see Sect. 1.4.3), and for not too large an anisotropy $\delta\epsilon/\epsilon$ the dynamic correlation function $g_1(t)$ can be written in a form very similar to (1.15).

1.3.4 Imaging with Diffusing-Wave Spectroscopy

Recent years have seen substantial progress in optical imaging "beyond the transport mean free path" (see e.g. [105] and references therein). Various techniques, such as interferometric detection of the weak unscattered coherent beam, time-resolved selection of early-arriving almost-unscattered photons, and measurements of photon density waves and diffuse photon intensities, have been applied in order to locate and eventually image objects which are buried several optical transport mean free paths deep inside the medium. In these techniques the optical contrast of the object with respect to the turbid medium is due to enhanced transparency or enhanced absorption, both of which modify the spatial distribution of the diffuse light intensity. The object basically acts as a source or a sink for diffusing photons, and therefore generates a glow or shadow respectively on the sample surface. The glow or shadow is less in amplitude but larger in size for deeply buried objects than for objects near the surface, because of the diffusive spread of photons from the object to the surface. This allows one to localize the object. The spatial resolution degrades roughly linearly with the distance of the object from the surface [106].

The DWS principle can also be used to image or locate objects which have *dynamic* contrast because of some motion with respect to the surrounding medium. While in this case the average scattered intensity does not necessarily depend on the position at the sample surface, the temporal fluctuations of it – as seen in $g_1(t)$ – do. This idea was suggested by work on speckle tomography [107], which pointed out the fact that if scatterers are moved even a small distance, the corresponding changes of the speckle pattern are most pronounced in the surface region closest to these scatterers. Boas et al. [89, 108] have reported images of a spherical cavity containing a colloidal suspension in Brownian motion (with $\ell^* = 1.5$ mm), located 0.75 diameters below the surface of a solid multiple-scattering medium (with $\ell^* = 2.2$ mm). In this case, there was contrast both in the scattering and in the dynamics. Heckmeier et al. [109–112] have performed experiments on objects having different types of purely dynamic contrast (i.e. identical ℓ^* values inside and outside the object). Position-dependent $g_1(t)$ measurements [109] from a capillary containing a flowing colloidal suspension embedded in the same suspension undergoing Brownian motion revealed that the flow rate, depth and in-plane location of the object can be obtained and $g_1(t)$ is in excellent agreement with simple photon diffusion theory [110]. The dynamic contrast has a maximum for a well-defined correlation time t. This is because $g_1(t)$ at very short times is dominated by Brownian motion (which is of course identical inside and outside the object) as compared to flow, while at very long times only scattering paths too short to sense the embedded object contribute to $g_1(t)$.

It is also possible to obtain dynamic contrast between Brownian particles having different sizes [110]. In the particularly sensitive situation in which the $g_1(t)$ measurement was made on a dark spot of the static speckle of a solid background medium containing a dynamic inclusion undergoing Brownian motion or flow, objects could be located as deep as five diameters and more than $30\ell^*$ inside the medium [111]. Finally, the probability distribution of the scattered intensity sampled at long times, rather than the full time dependence of $g_1(t)$, can be used for imaging purposes [112].

1.3.5 Perspectives

The above principles are expected to turn out useful in many applications, in particular in biomedical sciences. This may be illustrated by a recent experiment [208] on superficial burns of animal tissues, where indications about the depth of burn could be obtained from the analysis of the temporal decay of $g_1(t)$; the superficially burned layer of tissue behaves like a solid, while the nonburned tissue below generates time-dependent speckle fluctuations due to blood flow.

Diffusing-wave spectroscopy has become a very useful tool to probe dynamical properties of multiple-scattering media of various kinds. Studies on calibrated colloidal suspensions have borne out its potential to investigate fundamental problems in the physics of fluids, as illustrated for instance by the observation [94–96] of the short-time motion of spherical particles governed by hydrodynamic interactions. Many novel contributions of DWS to diverse problems in statistical physics are expected, primarily because of the wide range of time and distance scales covered. The quantitative understanding of DWS allows one to tackle more complex systems now. Foams, sand, liquid crystals, emulsions and polymer gels doped with scattering particles have been mentioned briefly, and many more applications are foreseen, particularly important perhaps for the quality control of food, cosmetics and paints. Multiple-scattering imaging and remote sensing of buried objects in motion may evolve into a versatile tool of particular interest in medical applications, given the relatively low optical extinction of biological tissue in the near infrared and the possibility to select particular objects spectroscopically. Examples include blood vessels, coagulates and dye-stained tumors. Such applications will be complementary to and, with the availability of low-cost sources and detectors of light, substantially cheaper than current NMR or X-ray imaging techniques.

1.4 Coherent Beam, Diffuse Beam and Speckles: A New View

Studies in the last 10 years have substantially modified the old picture of multiple scattering of light. The most dramatic revolution was undoubtedly in 1985 when the first experimental reports of coherent backscattering came in. This phenomenon is now successfully explained in terms of constructive interference between two waves propagating in opposite directions. New phenomena have also been found for the coherent beam and the speckles.

1.4.1 Diffuse Beam: Coherent Backscattering and Localization

Roger Maynard, Bart van Tiggelen, Georg Maret, Ad Lagendijk and Diederik Wiersma

On the basis of reciprocity, interference between two opposite paths can be argued to be constructive in the backscattering direction of, for instance, a slab geometry, and *exactly* as large as the conventional diffuse background calculated from (1.8). At backscattering, the equation of radiative transfer is thus 100 % wrong! As always, the width of an interference effect is roughly given by the wavelength divided by the typical distance between two typical points of scattering, in this case the mean free path, giving $\Delta\theta \approx 1/k\ell$ [113]. One can still argue as to what mean free path should be used here: the transport or the scattering mean free path. Although a physical argument favors the first (recall Fig. 1.2), a rigorous confirmation for anisotropic scatterers (for which both mean free paths differ) has only been given recently [114, 115]. Thus

$$\Delta\theta \approx \frac{1}{k\ell^*}.\tag{1.17}$$

The smallness of $1/k\ell^*$ in typical experiments probably explains why the serendipitous discovery of coherent backscattering was unlikely (Fig. 1.4).

Coherent backscattering has been investigated in a variety of circumstances. The general reciprocity relation that can be written down between the transition matrix (relating the incoming and outgoing electric fields of the light) of any event, D, and that for the same event in the opposite sequence, R, placed in a magnetic field $\mathbf{B}_0$, is [22]

$$D(\sigma, \mathbf{k} \to \sigma', \mathbf{k}' \,|\, \mathbf{B}_0) = R(\sigma', \mathbf{k}' \to \sigma, \mathbf{k} \,|\, -\mathbf{B}_0),\tag{1.18}$$

where $\sigma\ (=\pm)$ indicates the two possible states of circular polarization. In the absence of a magnetic field one can verify that $D(\sigma, \mathbf{k} \to \sigma, -\mathbf{k}) = R(\sigma', \mathbf{k} \to \sigma, -\mathbf{k})$. This means that for the diagonal channel $\sigma = \sigma'$ the inverse scattering sequence has the same scattering amplitude as, and therefore interferes constructively with, its opposite partner. More precisely,

$$|R + D|^2 = |R|^2 + |D|^2 + 2\,\mathrm{Re}\,RD^* = 2(|R|^2 + |D|^2)$$

at backscattering. This argument leads to the famous and apparently universal factor of two for the diagonal polarization channel. Absorption is allowed and therefore does not change this conclusion. Reciprocity does not inform us about the off-diagonal helicity channel. Experiments [116] and calculations [117–119] give a value of only 1.12 for this channel.

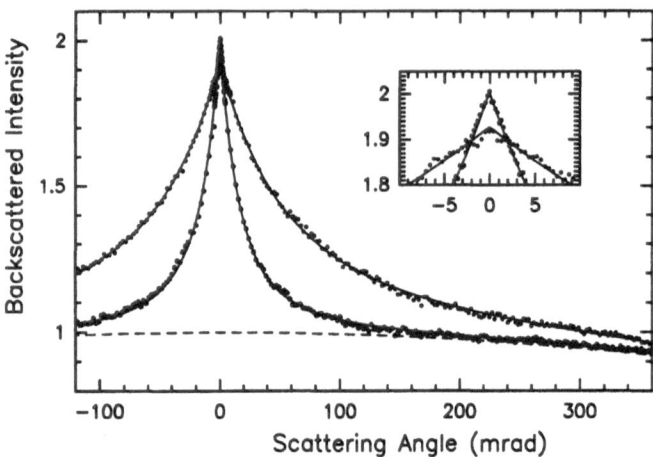

Fig. 1.4. A measurement of optical coherent backscattering from a dielectric random medium, performed with circularly polarized light with a vacuum wavelength of 633 nm. The backscattered intensity is plotted against the scattering angle, where zero corresponds to exact backscattering. The sample is BaSO$_4$ with a transport mean free path of roughly 2.1 μm. The high quality of this measurement was made possible by a new technique [122]. The measurement clearly shows the nonanalytic cusp in the center caused by extreme long-range diffusion [reprinted from Wiersma, Van Albada, van Tiggelen and Lagendijk, Phys. Rev. Lett. **74**, 4193 (1995), with permission from the American Physical Society]

The relation $D = R$ at backscattering can no longer be obtained from (1.18) once a magnetic field is present. The Cauchy inequality indicates that the factor of two can only shrink. In Fig. 1.5 we demonstrate coherent backscattering curves obtained by Erbacher, Lenke and Maret in an external magnetic field [21]. In a magnetic field the electric polarization vector is subject to Faraday rotation. This experiment proves that Faraday rotation kills the constructive interference, as allowed by reciprocity. These observations are in rather good agreement with calculations and numerical simulations carried out by Martinez and Maynard [22] (Fig. 1.6). The decrease of the enhancement factor seems to be a universal function of $VB\ell^*$, where V is the Verdet constant of the medium describing the rotation angle per tesla, per meter. So far, the equi-intensity lines of the scattering cone have been found to be circular, as expected if the direction of the magnetic field is along the backscattering direction. In order to verify a recent prediction [23] that these lines become elliptical because the diffusion tensor becomes a nondiagonal tensor in a magnetic field, one has to change the direction of the magnetic field.

Coherent backscattering has also been studied in relation to gain (using scatterers containing dye) [19], as well as in relation to internal reflection on the boundaries of the medium [120, 121]. Neither one of these mechanisms is believed to break the basic reciprocity argument leading to the factor of

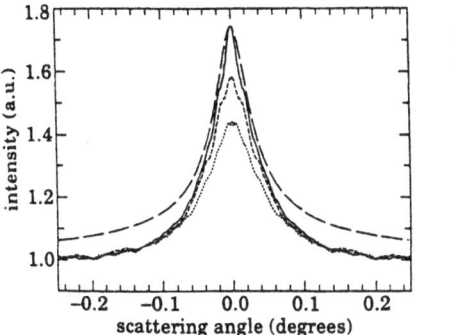
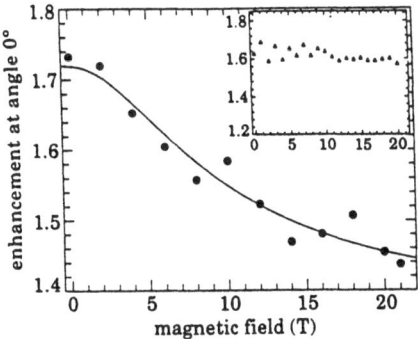

Fig. 1.5. Coherent backscattering in an external magnetic field. *Left:* angular dependence of the scattered light intensity in the vicinity of backscattering and for circular polarization, as obtained by azimuthally averaging video pictures around the peak position, and normalized to the flat background at 0.33°, in different magnetic fields: 0 T (*solid line*), 10 T (*short-dashed line*) and 21 T (*dotted line*). The *long-dashed line* corresponds to a fit to theory. The sample – 40 vol-% FR5 (a rare-earth-doped paramagnetic Faraday rotator glass), milled to a powder – had a length $L = 2$ mm and a transport mean free path $\ell^* \approx 70\,\mu$m. *Right:* magnetic-field dependence of the backscattered light intensity at exact backscattering of the same sample. The continuous line is a fit to theory. The inset shows a reference sample with negligible Faraday rotation but with same cone width [reprinted from Erbacher, Lenke and Maret, Europhys. Lett. **21**, 55 (1993), with permission from Les Editions de Physique, France]

two. The angular shapes may change however, because in both cases events involving long scattering paths will be favored, leading to a narrowing of the peak. In Fig. 1.7 we show measurements of the cone carried out in Amsterdam with gain [19]. These measurements confirm the picture above. The study of gain in combination with multiple scattering comes into the picture because it may offer the possibility of a random laser once the gain exceeds a critical threshold [123–125].

Does reciprocity really lead to an enhancement of exactly two in the helicity-conserving channel at backscattering? Indeed, if one applies the equation of radiative transfer and adds the coherent backscattering phenomenon required by reciprocity, this factor of two follows. However, this procedure violates flux conservation since the flux of the cone, though small and of order $1/k\ell$, is added ad hoc and is energetically not accounted for. At present a practical transport equation obeying both reciprocity and flux conservation is not available. A recent experiment done by Wiersma et al. in Amsterdam demonstrated for the first time the need for such an equation [36]. This experiment used samples with a value of $k\ell^* \approx 5$, and showed an enhancement factor lower than the "holy" factor of two predicted by the conventional approach for the helicity-conserving channel of circularly polarized light. The measured enhancement factor turned out to depend on the density of the scatterers, which excludes the alternative explanation of nonsphericity of the

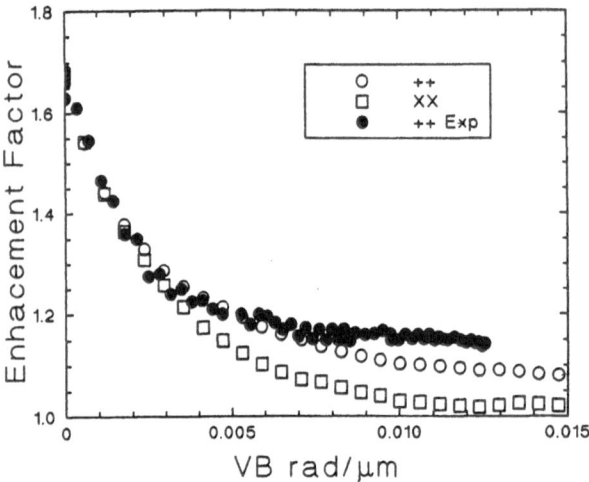

Fig. 1.6. Numerical Monte Carlo simulation of coherent backscattering in the presence of Faraday rotation. The experimental points were obtained by Erbacher, Lenke and Maret [21]. The numerical simulations were carried out by Martinez and Maynard [22] and show the coherent-backscattering enhancement factor in the ++ circular-polarization channel and in the xx linear-polarization channel. The input parameters for the numerical simulation were: $\lambda_0 = 0.4579$ µm, Mie particle radius $a = 0.1$ µm, and indices of refraction of 1.45 for the particle and 1.65 for the surrounding medium. The Verdet constant in the medium was $V = 1571$ rad/(mT) and was assumed to be zero inside the scatterers. In the experiment the Verdet constant is known to depend on the magnetic field. Experimentally, $L/\ell^* = 500$ [reprinted from Martinez and Maynard, Phys. Rev. B **50**, 3714 (1994), with permission from the American Physical Society]

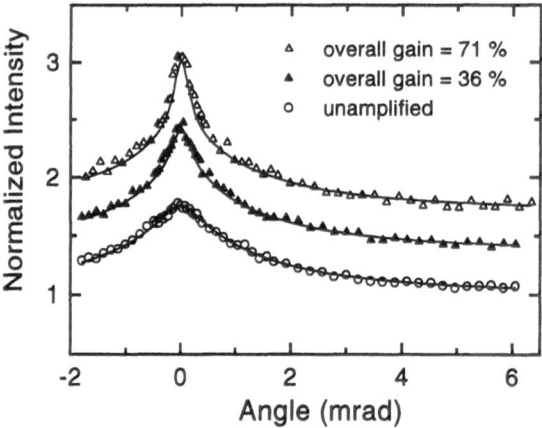

Fig. 1.7. Coherent backscattering in a medium with gain. As the gain of the medium increases, long paths of the photons achieve more weight and the coherent backscattering peak narrows [reprinted from Wiersma, Van Albada and Lagendijk, Phys. Rev. Lett. **75**, 1739 (1995), with permission from the American Physical Society]

particles. Only particles with rotational symmetry have a vanishing single-scattering signal in the helicity-conserving reflection channel. Since, in single scattering, only particles in the skin layer one mean free path in thickness contribute, the signal would be proportional to the particle density times the mean free path, i.e. independent of density. The experimental results could be reasonably explained by recurrent scattering of light between two particles, which can be seen as a sort of "super single scattering" from particles without rotational symmetry, leading to a density-dependent cross-section. This explanation also restores energy conservation to second order in the particle density [126].

The combination of flux conservation and reciprocity leads to a self-consistency problem that touches the heart of microscopic theories of strong localization [127]. The point is that interference not only modifies reflection in backscattering, but also changes the path-length distribution *inside* the system, and thus the background intensity. As a matter of fact the whole concept of "path-length distribution", a widely used term to refer to the contribution of "photon paths" of a particular length to the measured intensity, breaks down when interference is allowed. This complicated problem has so far only been considered in the diffusion approximation. In this theory one calculates the interference contributions to the diffusion constant (1.4), thereby requiring reciprocity and flux conservation. The result is, in three dimensions [127],

$$\frac{1}{D} = \frac{1}{D_{\mathrm{B}}} + \frac{1}{D}\frac{1}{4\pi k^2 \ell} \int_{q_{\min}}^{q_{\max}} \mathrm{d}^3\mathbf{q}\,\frac{1}{q^2}\,. \tag{1.19}$$

D_{B} is the diffusion constant without interference, $q_{\min}$ is a lower cutoff related to the finite sample size and $q_{\max} \approx \pi/\ell$ is an upper cutoff denoting a lower length at which the diffusion approximation breaks down. Although this theory is far from rigorous, it can be shown to agree with the (phenomenological) scaling theory of localization [128]. Moreover, it predicts strong localization, here defined as $D = 0$, to occur in an *infinite* medium when

$$k\ell \approx 1\,; \tag{1.20}$$

this is known as the Ioffe–Regel criterion and was first proposed in the sixties by Mott as a criterion for localization. In a *finite* medium Anderson localization leads to a geometry-dependent diffusion constant and finally to an (ensemble-averaged) transmission that decays exponentially with the size of the medium, not to be confused with the almost trivial exponential decay of the coherent part in (1.6). According to this theory strong-localization phenomena will be demolished in a magnetic field [129]. Unfortunately, this does not agree with exact calculations of the Anderson model with a magnetic field [130] or with random-matrix theory [131]. Furthermore, the observation of light localization in more than one dimension has turned out to be more difficult than suggested by the Ioffe–Regel criterion. At the time of writing, there are reports for microwave localization in two [34] and three [33] dimensions.

1.4.2 Diffuse Beam: Photonic Hall Effect

Bart van Tiggelen and Geert Rikken

In the old picture of multiple scattering of light the diffusion tensor was always seen to be isotropic. Recent work addresses anisotropic components in the diffusion tensor. Two different systems are being actively studied that are now known to exhibit anisotropic light diffusion: magnetoactive media and nematic liquid crystals.

The photonic Hall effect has recently been predicted theoretically [37]. It consists of a diffuse current which is perpendicular to both the magnetic field and the density gradient. In this case, (1.4) is modified to

$$J = -D_0 \nabla \Phi - D_H \mathbf{B} \times \nabla \Phi. \tag{1.21}$$

The medium under consideration is a random medium with small Rayleigh particles that are magnetoactive, meaning that they are made from a material in which light would exhibit Faraday rotation in the bulk. The constant D_H in (1.21) is, quite surprisingly and unlike D_0, *independent* of scatterer density. In addition, it is linearly proportional to the Verdet constant V of the material. This is interesting because – as a rule of thumb – paramagnetic materials have a negative Verdet constant, whereas diamagnetic materials normally have a positive Verdet constant. This means that the direction of the current will not be the same for dia- and paramagnetic materials. In addition, for paramagnetic materials the Verdet constant is inversely proportional to the temperature. It can be demonstrated that the Hall component of the light current satisfies C, P and T symmetry (Table 1.2), as well as the Onsager relation $\mathbf{D}(\mathbf{B}) = \mathbf{D}^t(-\mathbf{B})$. The T symmetry is especially surprising, recalling that one is considering *averaged* quantities. For this reason the first term in (1.21) does not obey T symmetry.

Table 1.2. Transformation under parity (P), time reversal (T) and charge conjugation (C) of the relation for the transverse photon current $\mathbf{S}_\perp \sim V\mathbf{B_0} \times \nabla \Phi$. Note that the Verdet constant V has the same symmetry as "charge" and plays more or less the role of charge for the light

Symmetry operation	$\mathbf{S} = \mathbf{E} \times \mathbf{H}$	$\nabla \Phi$	$\mathbf{B_0}$	V
P	−	−	+	+
T	−	+	−	+
C	+	+	−	−

Experiments using a 40 Hz oscillating field and synchronous detection verified the theoretical predictions [38]. In Fig. 1.8 we show the photonic Hall current, measured in a cylinder filled with (paramagnetic) CeF_3 in glycerol. Also confirmed have been the $1/T$ temperature dependence, as well as the

CeF₃, 77 K

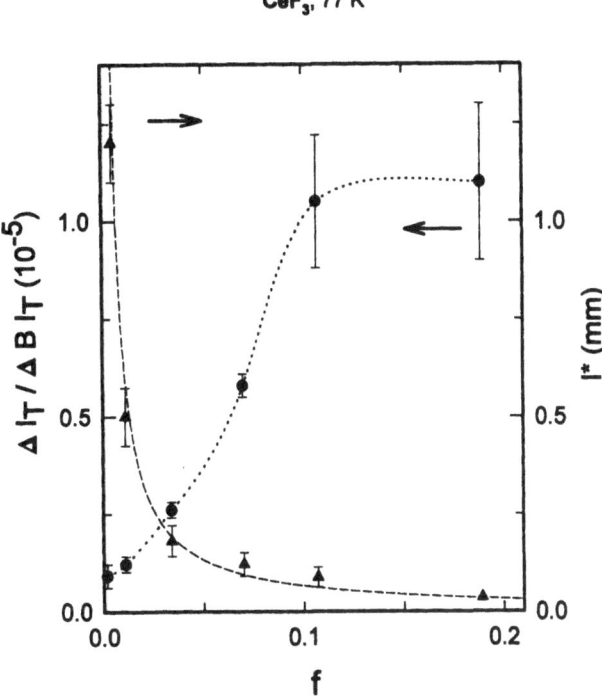

Fig. 1.8. Observation of the "photonic Hall effect". The figure shows the transversely detected light current through ΔI_T the walls of a cylinder (length $L = 1.3$ mm, radius $R = 0.5$ mm), perpendicular to the direction of the magnetic field and the main diffuse current. The signal has been normalized to the transverse losses I_T detected in the absence of the magnetic field and the magnetic field ΔB. The cylinder contains a mixture of glycerol and strongly magnetoactive 2 μm CeF₃ powder ($T = 77$ K). The normalized magneto-transverse current is plotted against the volume fraction f of the CeF₃ powder. The transport mean free path ℓ^* (right axis) has also been determined from transmission experiments. For $f > 10\%$ the mean free path is smaller than the cylinder dimensions [reprinted from Rikken and van Tiggelen, Nature **381**, 54 (1996), with permission from Macmillan Magazines Ltd.]

linearity with the applied field. This experiment constitutes the first experimental confirmation that light diffusion can be anisotropic. At the time of writing, the longitudinal photonic magnetoresistance, the photonic equivalent of the electronic magnetoresistance, has also been confirmed experimentally [132]. This effect is observable in transmission in a slab geometry and is proportional to $(VB\ell^*)^2$, as predicted in [23].

1.4.3 Diffuse Beam: Optics in Nematic Liquid Crystals

Bart van Tiggelen and Roger Maynard

The second example of anisotropic light diffusion occurs in nematic liquid crystals. Two complications show up in these systems. First, the nematic liquid crystals have uniaxial symmetry, with one optical axis n (the dielectric tensor is given by $\varepsilon_{ij} = \varepsilon_\perp \delta_{ij} + \varepsilon_a n_i n_j$). For light propagation this gives rise to two modes with different dispersion relations (birefringence), of which one is not spherical, and for that reason the latter is called "extraordinary". Secondly, scattering of light is caused by thermal fluctuations, and not by scattering from the individual molecules. Owing to the rather complex free energy that describes these fluctuations, the scattering process is highly unusual, with significant polarization effects. Uniaxial symmetry constrains the diffusion tensor to be of the form [79–84]

$$D_{ij} = D_\perp \delta_{ij} + (D_\parallel - D_\perp) n_i n_j \,. \tag{1.22}$$

This implies two different diffusion constants, along and parallel to the optical axis. In Fig. 1.9 we show some theoretical predictions. An independent theoretical approach by Stark and Lubensky [81] is in agreement with these results. Multiple scattering with two modes having different speeds and polarizations has also become important for elastic waves, and its relevance to seismology will be discussed in Chap. 7. A recent experiment on the compound 5CB carried out by the Pennsylvania group [82] confirmed this diffuse anisotropy in agreement with theoretical predictions.

1.4.4 Coherent Beam

Ad Lagendijk and Bart van Tiggelen

The coherent beam is characterized by a complex effective-medium dielectric "constant", and at present any progress involves a more sophisticated calculation of this "constant". Unlike to the calculation of the diffuse beam, it is difficult to say a priori if a calculation is "good" or "bad". For the diffuse beam a calculation can be called "good" when it breaks neither flux conservation nor reciprocity (implying unfortunately that there exist very few "good" calculations for the diffuse beam). For the coherent beam, every calculation satisfies reciprocity (there is no interference in the coherent beam, only superposition) and no calculation obeys flux conservation, because this has already been destroyed by the averaging. The absence of symmetry arguments means that when you do a sophisticated calculation, you have to compare it with experiment to see if it is correct. With the advent of fast computers, very sophisticated effective-medium theories have been developed [55, 133, 134], and compare well to the experiments. (See also Sect. 1.3) Coherent-beam experiments have become popular again, with the recent method of heterodyne detection.

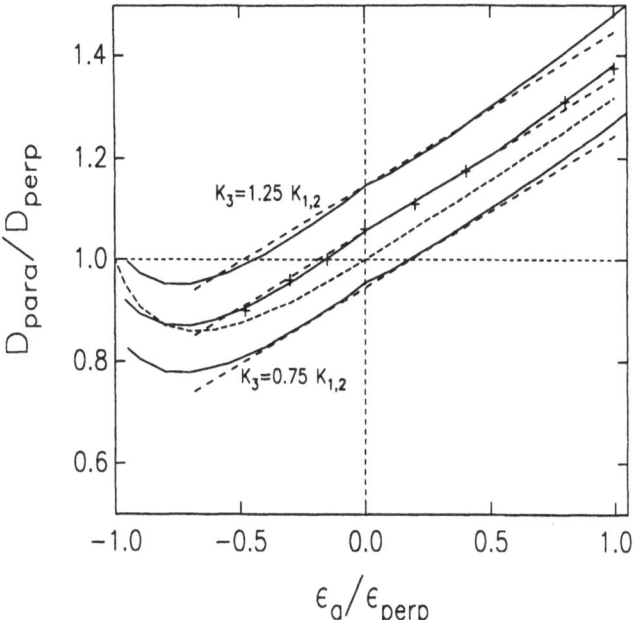

Fig. 1.9. Theoretical predictions for the diffusion anisotropy in a nematic liquid crystal as a function of the uniaxial dielectric anisotropy. The *middle solid line* corresponds to an isotropic free energy; the other two lines assume equal strengths of splay and twist distortion, but different bend strengths. The *fine dashed line* through the origin represents the kinematic anisotropy discussed in [79] and [83]. The points (*crosses*) are taken from Fig. 8 of [84]. The three parallel *dashed lines* denote the Taylor expansion given in (123) of [84]. The correlation length of the thermal fluctuations is fixed at 5 μm but a change would hardly affect this figure

Three different mechanisms can be identified that affect the complex dielectric constant: local-field effects, recurrent scattering and particle correlations. The first will not be discussed here. It concerns the denominator in (1.7). Local-field effects were already part of the "old view" and introduced by Lorentz. However, they are receiving new attention because they are highly relevant and somewhat controversial in the description of the spontaneous emission coefficient in inhomogeneous materials [135, 136].

Other density modifications to the complex dielectric constant come from recurrent scattering between two or more particles. These modifications establish an interesting link with induced dipole–dipole coupling [137]. This interaction is known to be due to recurrent exchange of light waves (either classical or quantized), which is essentially the same process as in the dielectric constant, but now seen in the physical world of the particles.

Recurrent scattering effects are sensitive to the statistics of the scatterers. This notion has recently seen an interesting application in degenerate Bose gases. By accurately measuring the index of refraction of Bose–Einstein

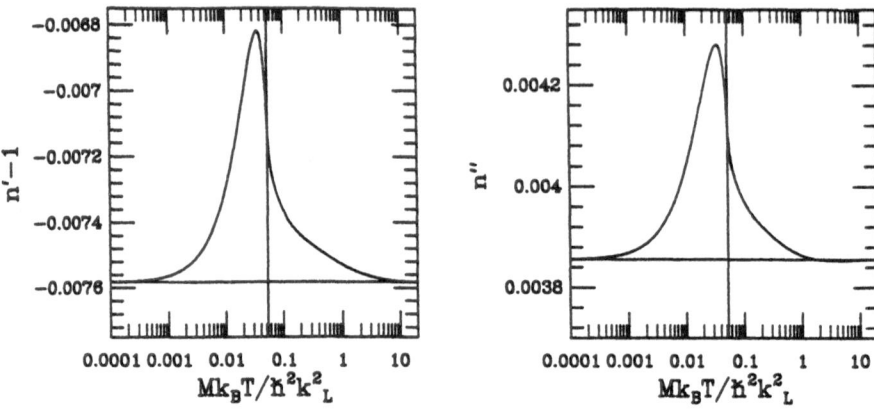

Fig. 1.10. Real part (n') and imaginary part (n'') of the refractive index of an ideal (noninteracting) Bose gas (*curves*) and of a gas of distinguishable particles (*horizontal lines*), as function of the temperature. The atom–laser detuning is $\delta = \Gamma$ (where Γ is the natural line width) and the atomic density is $\rho_0 = 0.5/\lambda_L^3$, where $\lambda_L = 2\pi/k_L$ is the wavelength of the laser. The vertical line indicates the critical temperature where Bose–Einstein condensation occurs for the ideal Bose gas [reprinted from Morice, Castin and Dalibard, Phys. Rev. A **51**, 3896 (1995), with the kind permission of the authors and the American Physical Society]

condensates, recurrent-scattering modifications may enable one to deduce quantum pair correlations in a noninvasive way. In Fig. 1.10 we show the predicted complex index of refraction of a Bose gas, compared to a classical gas, on the basis of calculations by Morice, Castin and Dalibard [138].

Cluster effects clearly become relevant if the scatterer density is high. One way to incorporate cluster effects in any multiple-scattering equation is to consider the interference between the scattered fields of correlated particles as described by the (static) structure factor $S(\mathbf{q})$. This suggests that the differential scattering cross-section of a cluster of particles is given by the cross-section of an individual particle times this correlation factor [139],

$$\frac{d\sigma}{d\Omega}\left(\mathbf{k} \to \mathbf{k}'\right) = S\left(\mathbf{k} - \mathbf{k}'\right)\left(\frac{d\sigma}{d\Omega}\right)_0\left(\mathbf{k} \to \mathbf{k}'\right) .$$

If excluded volume is present, this expression tends to increase the scattering mean free path ℓ of the coherent beam. It also modifies the angular profile, and thus the transport mean free path of the coherent beam. This expression is often used, despite the fact that it ignores both local-field effects and recurrent scattering. Within this approximation, a great advantage is that a measurement of ℓ^*/ℓ gives direct information on the structure factor, and hence about particle correlations. This notion has already seen interesting applications in colloidal media [140].

1.4.5 Speckles

Bart van Tiggelen

A lot of recent research has been devoted to speckle phenomena going beyond the Siegert relation (1.10) and Rayleigh statistics. We shall only briefly indicate the recent state of the art, and refer to other review papers for details [141]. This domain is now in a stage of rapid development, thanks in particular to the application of random-matrix theory.

The Siegert relation (1.10) is mathematically the "second-order cumulant" expression of the correlator $\langle \psi_a \psi_a^* \psi_b \psi_b^* \rangle$. The fourth-order cumulant has been neglected. Recent studies have established the more general expression [142] [143],

$$\frac{\langle I_{ab} I_{a'b'} \rangle}{\langle I_{ab} \rangle \langle I_{a'b'} \rangle} = 1 + C^1_{aba'b'} + \frac{1}{g} C^2_{aba'b'} + \frac{1}{g^2} C^3_{aba'b'} . \tag{1.23}$$

Here, I_{ab} denotes the transmission coefficient from channel a to channel b (think of a channel as a discretized version of an incident or outgoing directions) and g is a dimensionless quantity, often called the dimensionless conductance. This parameter originally showed up in the scaling theory for localization [128] and is the most important parameter in modern speckle theory. If N is the total number of channels in a system with typical length L, then g is N times the total transmission: $g \approx N\ell^*/L$. If the width of the same system is W, the total number of independent speckle spots is estimated to be $N \approx Wk^2$. If $W \approx L$ and $k\ell^* \gg 1$ it is easily verified that $g \gg 1$. Consequently, the corrections obtained in (1.23) are in principle very small. The C^1 term represents the familiar fluctuation characteristics derived in (1.10).

Calculations for transmission show that $C^2_{aba'b'}$ and $C^3_{aba'b'}$ decay very slowly if the separation between the modes becomes large [142] (the first algebraically and the second hardly at all, whereas C^1 decays exponentially). Asymptotically, they can thus nevertheless prevail. However, this is difficult to measure. Fortunately, the different contributions also exhibit different selection rules.

The C^1 term is only present when $|\mathbf{q}_a - \mathbf{q}_{a'}| = |\mathbf{q}_b - \mathbf{q}_{b'}| < 1/L \approx 0$. Note that this implies an interesting feature, namely that if the incident beam shifts by a certain amount, the speckle pattern in transmission will statistically shift by the same amount. This is called the memory effect, and has been observed experimentally [144]. If one sums over either all incident or all outgoing channels, the C^1 contribution becomes negligible with respect to the others.

The C^2 term prevails if one sums over either outgoing or incident channels. This makes the C^2 speckle relevant in "all-channel-in, one-channel-out" or "one-channel-in, all-channel-out" experiments. These experiments are not easy to carry out but some investigators have succeeded in confirming the existence of C^2 speckle in multiple scattering of light [145, 146].

The C^3 contribution hardly depends on the channels. This implies that when summing over both all incident and all outgoing channels this term will be dominant. For electrons it is easier to do an "all-channel-in, all-channel-out" experiment because electrons are automatically on the Fermi surface. Summing over all channels of the transmission matrix implies that one basically considers the conductance $G = \Sigma_{ab} I_{ab}$. The fluctuations predicted in this way,

$$\Delta G \approx \text{constant} \approx 1, \tag{1.24}$$

are known as "universal fluctuations". They are undoubtedly one of the most astonishing new features in multiple scattering of waves. For light the $C^3 contribution$ has so far never been measured. For the electronic conductance of mesoscopic samples, the existence of the fluctuations has been verified [147]. Using the high-order cumulants, it is possible to calculate modifications to the conventional speckle intensity distribution [148]. These modifications have also been observed [149]. Recently, some investigators have succeeded in calculating the "one-channel-in, all-channel-out" distribution function [150–152]. Again, these calculations have been confirmed experimentally with microwaves [153].

1.5 Multiple Scattering of Microwaves

Patrick Sebbah and Azriel Genack

Wave propagation in random media involves interference processes between different scattered partial waves. A careful understanding of the multiple-scattering regime of propagation requires one to go beyond the random-phase approximation. For instance, coherent backscattering is a consequence of the unavoidable constructive interference between waves propagating in opposite directions [15–17]. In the very-strong-scattering limit, Anderson localization can only be understood through a nonperturbational scheme where all interference processes between all multiply scattered waves are taken into account [154]. Mathematically, the field $E = Ae^{i\varphi}$ can be represented by an amplitude A and a phase φ. Most of the time, experimental optics deals only with intensities $|A|^2$. The phase is, however, certainly a key parameter in the understanding of multiple scattering of waves, but recent techniques to retrieve the phase from intensity measurements are still limited [155–157]. Microwaves constitute a family of electromagnetic waves that exhibit the same features as optical waves but at lower frequencies, allowing a direct field measurement and phase retrieval.

1.5.1 Experimental Approach

In our experiments, the random medium consists of randomly positioned polystyrene spheres with a diameter of 1.27 cm, placed in a cylindrical copper tube (in order to restrict transverse diffusion), at a volume filling fraction $f = 0.55$. The sample tube is rotated between successive measurements to produce new scatterer configurations. The microwave radiation is coupled to the medium via a horn at the input and detected in transmission via a 0.4 cm wire antenna placed at 0.5 cm from the output surface of the sample. We use an 8720C Hewlett-Packard network analyzer, which synthesizes the microwave radiation between 40 MHz and 26 GHz with a frequency increment of 625 kHz and performs a coherent detection of the transmitted signal. The input face has been chosen as our zero phase reference.

1.5.2 Correlation Functions in Space and Frequency

Random phasing of partial waves reaching the detector leads to intensity fluctuations and gives rise to a speckle pattern which is readily seen in optical experiments. Short-range intensity correlations occur on the scale of the phase coherence length in the medium, which is the inverse of the wave vector and roughly the size of a speckle spot. But local intensity fluctuations diffuse throughout the sample, giving rise to longer-range correlations [158], which means that the intensities in distant speckle spots are not statistically independent.

As has already been shown in (1.23), the normalized correlation function C_I between intensities at nearby frequencies ν and $\nu + \Delta\nu$ and/or nearby positions $\mathbf{r}$ and $\mathbf{r} + \Delta\mathbf{r}$ can be expressed as an expansion in the small parameter $1/g$ (where g is the dimensionless conductance and is a measure of the closeness to the localization threshold, which occurs when $g \sim 1$) [159, 160]:

$$C_I(\Delta\nu, \Delta r) = C_1 + C_2 = A_1 F_1(\Delta\nu) H_1(\Delta r) + A_2 F_2(\Delta\nu) H_2(\Delta r), \quad (1.25)$$

with

$$A_1 = 1 + \frac{2}{3g} + \frac{4}{15g^2} \text{ and } A_2 = \frac{2}{3g} + \frac{2}{15g^2}. \quad (1.26)$$

F_1, F_2 (and correspondingly H_1, H_2) are functional forms which describe the dependence of the correlation frequency shift (and correspondingly with position shift), and are equal to 1 for $\Delta\nu = 0$ (and correspondingly for $\Delta x = 0$). The C_1 term corresponds to short-range correlations and gives the size of the speckle spot, while the C_2 term is a measure of the long-range intensity correlation between two different speckle spots. There is also a C_3 term, which corresponds to infinite-range correlation, but this is hardly detectable in our measurements.

The $F_1 H_1$ term is equal to $|C_E|^2$, where C_E is the field correlation function, and corresponds to the field factorization approximation of the intensity correlation function, so that $C_I = |C_E|^2$ in the absence of long-range correlation [161]. It turns out that measurement of the field enables us to discriminate between short-range C_1 and long-range C_2 correlations. At $\Delta\nu = 0$ and $\Delta x = 0$, $C_I = A_1 + A_2 \sim 1 + 4/3g$, while $|C_E|^2 = 1$, giving a direct access to g.

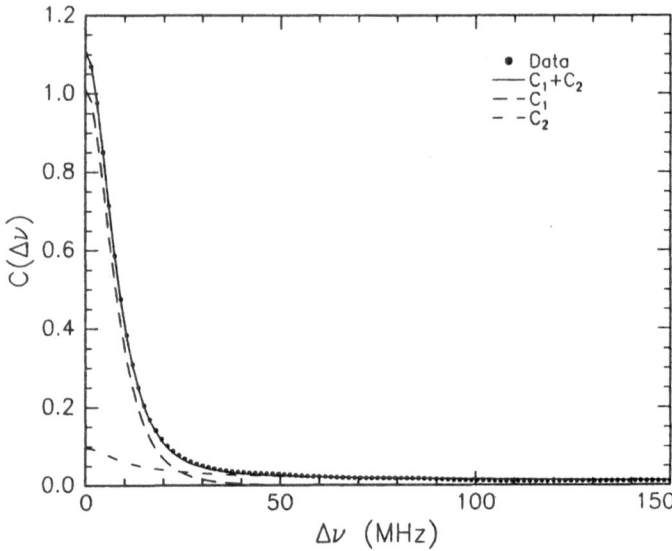

Fig. 1.11. Intensity correlation function with respect to frequency shift (*dots*). Theoretical fit of C_I by $C_1 + C_2$ (*solid line*). C_1 is a fit of $A_1 F_1 = A_1 |C_E|^2$ obtained from direct field measurement [reprinted from Genack, Polkosnick, Lisyansky and Sebbah [43], with permission from the Optical Society of America]

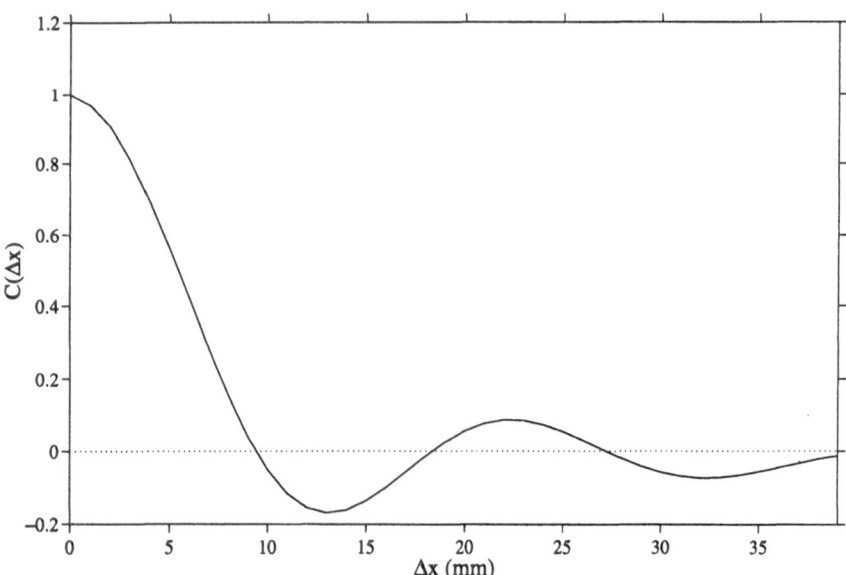

Fig. 1.12. Real part of the field correlation function with position shift

From the field measurements, we compute the correlation functions with respect to frequency shift ($\Delta x = 0$). In Fig. 1.11, the intensity correlation function with respect to frequency shift $C_I(\Delta\nu)$ is compared to the modulus squared of the field correlation function $|C_E(\Delta\nu)|^2$. From the value at $\Delta\nu = 0$, we estimate $g \sim 15$. Theoretical expressions for F_1 and F_2 [162] have been used to fit the data, with the diffusion constant D and the absorption time τ_a as fitting parameters.

Because microwaves have macroscopic wavelengths, fine scanning of the output speckle is possible. To get the position dependence of the correlation functions, we measured the field at different positions of the detector over a range of 4 cm with a regular spacing of 1 mm. The field correlation function with respect to position shift $\mathrm{Re}\,C_E(\Delta x)$ is shown in Fig. 1.12. It clearly demonstrates the presence of constant long-range correlations. The factorization approximation yields the following expression for H_1 [161]:

$$H_1 = \frac{\sin(k_0\Delta x)}{k_0\Delta x}\mathrm{e}^{-\Delta x/2\ell} \tag{1.27}$$

where ℓ is the mean free path. This simple expression yields a good fit, presented on the same figure.

1.5.3 Phase Statistics

Instead of the field, we may consider the phase of the detected radiation transmitted through the random medium. The phase measured by the instrument is modulo 2π. We have proposed [43, 163] to construct the cumulative phase $\rho(\nu)$ at frequency ν from the phase-determined modulo 2π. To do so, we follow the phase variation from low frequencies, where the phase approaches zero, up to ν, using frequency increments small enough for the changes of the phase to be less than π rad.

In free space or in a homogeneous medium, the cumulative phase is a measure of the phase accumulated by the wave during propagation. However, for propagation in multiple scattering media, the significance of the cumulative phase does not seem to be straightforward. The complex field at the detection point is the sum over a large number of partial waves with random phase emanating from the source, $E\mathrm{e}^{\mathrm{i}\varphi} = \sum p_\alpha \mathrm{e}^{\mathrm{i}\varphi_\alpha}$, where φ_α is the phase accumulated along the wave path α and p_α is its magnitude. As a consequence, a slight change in frequency or scatterer position will only slightly modify each φ_α, but can result in a dramatic and unpredictable change of the phasor sum. It is reasonable, first, to explore the statistical properties of the cumulative phase.

Following the procedure outlined above, we have constructed the cumulative phase for each of 581 sample configurations, between 6 and 26 GHz. The average $\langle\varphi\rangle$ is shown in Fig. 1.13. Two linear regimes are readily seen, which correspond to a ballistic regime from 0 to 10 GHz (where the wavelength is much larger than the scatterer diameter) and a diffusive regime from 15 GHz

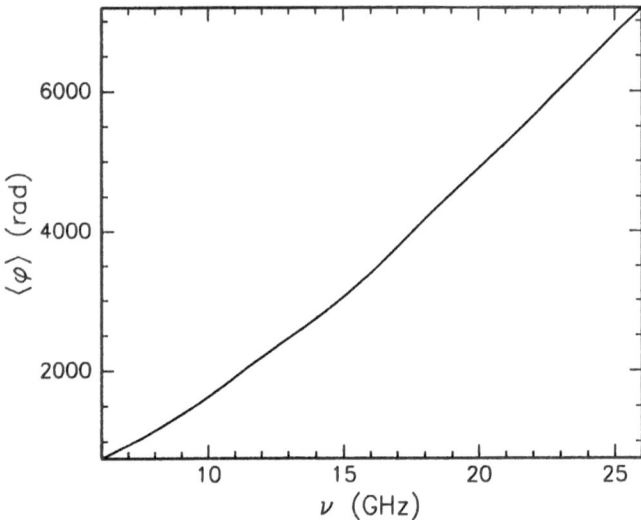

Fig. 1.13. Ensemble-averaged cumulative phase versus frequency [reprinted from Sebbah, Legrand, van Tiggelen and Genack, Phys. Rev. E **56**, 3619 (1997), with permission from the American Physical Society]

to 26 GHz (where the wave is multiply scattered). We studied the probability distribution of the cumulative phase, which we found to be a Gaussian at all frequencies, with a variance nearly equal to its average. These results can be interpreted in terms of short-range correlation of the phase derivative at all frequencies [163].

For a deeper understanding of the cumulative phase, we investigated its derivative with respect to frequency $d\varphi/d\nu$ and looked for connections with time-domain quantities. We showed both experimentally and theoretically [164] that the time for a pulse to cross the tube coincides exactly with the phase derivative, in the limit where the input pulse duration is very long compared to the inverse frequency increment. Figure 1.14 presents, in the same plot, the envelope of a $\Delta t = 10$ ns input pulse centered around $t = 0$, with carrier frequency $\nu_0 = 8.5$ GHz, and the envelope of the corresponding output signal $E_{\Delta t}(\nu_0, t)$. The travel time of this pulse through the sample is defined by

$$\tau_{\Delta t}(\nu_0) = \frac{\int t|E_{\Delta t}(\nu_0, t)|^2 dt}{\int |E_{\Delta t}(\nu_0, t)|^2 dt}. \tag{1.28}$$

When the pulse duration is 200 ns, this time delay across a particular config-uration is exactly the phase derivative at ν_0, $d\varphi/d\nu(\nu_0)$, which is represented in Fig. 1.15 for ν_0 between 8 and 9 GHz. However, even if this result is true for a given configuration, it is difficult to use because the infinite-pulse limit has no physical reality. On the other hand, for a finite square pulse, there is no equality between the dwell time $\tau_{\Delta t}(\nu_0)$ and the phase variation

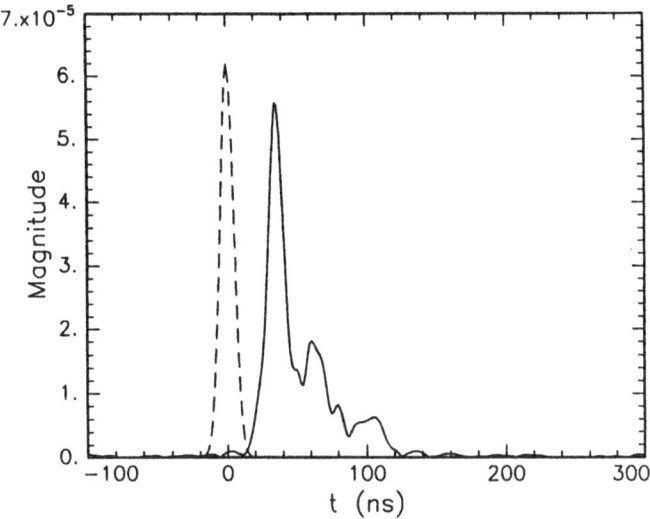

Fig. 1.14. A 10 ns input pulse (*dashed line*) and the corresponding output pulse (*solid line*) after transmission through the random sample (for one configuration)

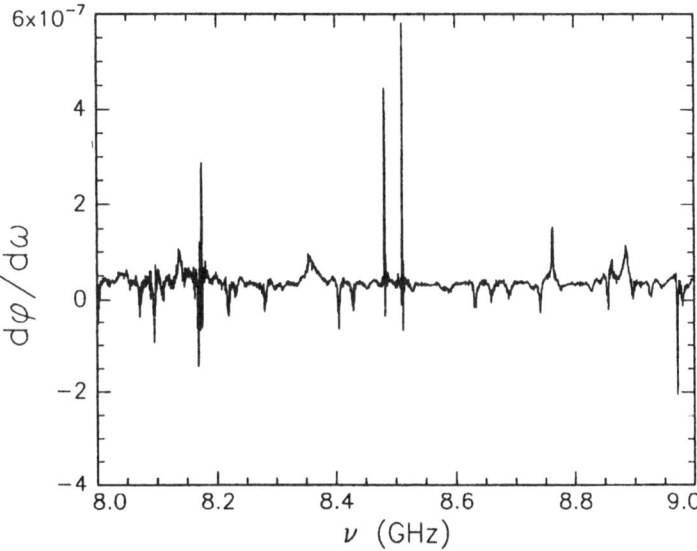

Fig. 1.15. Phase derivative (indistinguishable from the measured dwell time) for frequencies between 8 and 9 GHz (for one configuration) [reprinted from Sebbah, Legrand and Genack [164], with permission from the Optical Society of America]

$\Delta\varphi/\Delta\nu(\nu_0)$, where $\Delta\nu = 1/\Delta t$. However, we found the equality to be true for the ensemble average, i.e. $\langle\tau_{\Delta t}(\nu_0)\rangle = \langle\Delta\varphi/\Delta\nu(\nu_0)\rangle$.

The linear increase of $\langle \varphi \rangle$ in Fig. 1.13 can be directly interpreted in term of a constant average passage time for the wave. This linear behavior can be understood by considering the dwell time in our absorbing sample, whose length L exceeds the absorption length L_a, where $L_a = \sqrt{D\tau_a}$ and D is the diffusion constant. In this case, the average transit time is proportional to the product of the transit time over one absorption length, L_a^2/D, and the number of absorption lengths in the sample, L/L_a. Thus, $\langle \tau \rangle \sim LL_a/D \sim L\sqrt{\tau_a/D}$ and is constant with frequency if τ_a/D is also constant. Indeed, previous measurements on this sample have shown that τ_a and D are proportional over a broad frequency range [165]. Thus the linear behavior of $\langle \varphi \rangle$ with frequency is a consequence of the properties of this particular sample.

We are now investigating further to find a direct relationship between the photon dwell time in a given sample and the phase derivative. A longer sojourn time associated with a peak in the phase derivative is related to a resonance in the sample. The density of states within the sample is proportional to the sum over all input channels of the dwell time in each of these channels [166–168]. Measurements of the cumulative phase should thus yield the density of states in random media.

2. Optical Medical Diagnostics and Imaging

Jean Virmont and Guy Ledanois

In the first part of this chapter we review various approaches to optical medical imaging which have seen recently, significant progress and some others which seem within reach and are highly desirable. In the second part we present the optical imaging of large human organs such as the breast or even the head.

2.1 Overview of Optical Imaging in Tissues

When a slice of human tissue is illuminated by a light beam – of visible, infrared (IR) or ultraviolet (UV) wavelength – it behaves as a scattering and absorbing medium. A narrow collimated beam (e.g. a laser beam) incident on a tissue slice is partially transmitted, its output intensity decreasing exponentially with the tissue thickness, with a characteristic (scattering) mean free path of typically 50 µm. A variable fraction of the incident light disappears by absorption, while the rest emerges from both faces of the tissue layer, at positions and with directions away from those of the incident beam. These processes are believed to be well described quantitatively by standard radiative transfer, using a scattering coefficient μ_s, a phase function $P(\theta)$ describing the angular distribution of scattered photons, an absorption coefficient μ_a and an optical index n that provides refraction at the interfaces.

The physical origin of the scattering resides in the numerous small-scale variations of the optical refractive index between the aqueous, lipid and protein components of the cells ($n \approx 1.33$, 1.46 and 1.51 respectively). The wavelength dependence of μ_s in the visible range is typically $1/\lambda$, in sharp contrast to Rayleigh scattering, which scales as $1/\lambda^4$. This means that whereas small scatterers are involved in the Earth's atmosphere, for example, the dominant scatterers in tissues are larger than the wavelength (one micron or more). The origin of absorption resides in the various cellular chromophores, with a minimum value in the near IR (0.7–1.2 µm). So, μ_a has a broad range of variation, from 0.02 cm^{-1} in the IR to 100 cm^{-1} in the UV. IR radiation is used for maximum transmission, and UV for fluorescence excitation. Note that while many experiments have confirmed the applicability of radiative transfer in tissues, some papers have reported discrepancies [169, 170].

For modeling, an important quantity is the effective scattering coefficient $\mu'_s = \mu_s \times (1 - g)$, where $g = \langle \cos \theta \rangle$. The lengths $\lambda_s = \mu_s^{-1}$ and $\lambda'_s = \mu_s'^{-1}$ are called the scattering mean free path ℓ and the transport mean free path elsewhere in this book. The latter is the distance over which light loses memory of its initial direction. The scattering is usually forward peaked, with $g \approx 0.9$ and giving the large value $\lambda'_s \approx 0.5$ mm.

For source–detector distances smaller than a few times λ'_s, modeling requires solving the radiative transport equation, and is usually carried out using the Monte Carlo method, which can cope with 3D media. For larger distances, the diffusion approximation applies, which allows one to write down various analytical expressions [171]. Precise validity conditions for these have been established [172].

Measuring scattering and absorption coefficients is not an easy task. Their accurate evaluation requires several experiments, involving thin layers if μ_s, $P(\theta)$ and g are desired, and thick layers for μ_a and μ'_s. Discussion of the many difficulties that severely limit the accuracy of in vivo measurements, particularly for μ_a, is beyond the scope of this short review. Data are both scarce and often contradictory, by up to factors of 10! In spite of much published work, further identification of simple, efficient but accurate measurement procedures is needed. Systematic studies are highly desirable, particularly for layered tissues such as skin, and inhomogeneous tissues such as breast and brain.

In this contribution, we choose to stick to a clinical point of view. We first summarize the state of the art for some applications of optics to medical diagnostics, particularly by means of structural or functional imaging. We insist on medical relevance, on recent progress and on issues to be solved.

2.1.1 Microscopy

Apart from the doctor's eye, the most basic optical tool in medicine is the microscope, which is of fundamental importance in anatomy and pathology for detecting anomalous cells. When looking at a tissue with a standard microscope, scattering is minimized by cutting a thin slice, while the absorption contrast can be enhanced by using specific colorants. A relatively recent development is 3D imaging with a confocal microscope (CFM). The incident light is tightly focused inside the tissue layer, and backscattered light is collected in such a way that only the fraction coming from the focal point reaches the detector. The focal point is scanned mechanically in three directions. This provides 3D imaging with remarkable resolution, giving access to submicron details of the cell structure.

Operated in reflection, the CFM has been adapted to look into superficial layers of thick tissues, particularly into skin in vivo. Compared to the standard CFM, increased penetration is demanded, at the price of a somewhat reduced spatial resolution. Images have been obtained from the first 100–150 μm of skin and epithelium, with a resolution of a few microns [173–175].

The image contrast of deeper layers seems to be washed out by scattering. This has been explained by Monte Carlo simulations, which follow the light rays both in the microscope and in the scattering medium. Stray light due to backscattering from shallow regions overwhelms images produced by objects situated deeper than 3–5 scattering mean free paths, because the light going to and coming back from these objects is exponentially attenuated [176, 177]. The exact penetration depth depends somewhat on the microscope aperture, pinhole diameter, scattering anisotropy, etc.

A closely related domain that has recently seen spectacular progress is the 3D imaging of the retina. Rather than adapting the CFM, a method using a collimated beam has been developed. Rather than using the time of flight to obtain longitudinal resolution, optical-coherence tomography (OCT) uses the coherence time of the incident light in combination with interferometric detection. A dynamic larger than 100 dB and a longitudinal resolution of 15 μm have been obtained, providing 2D retinal images, resolved in depth and one direction along the surface, with excellent contrast. Thanks to the stratified nature of the retina, observation of morphological anomalies that are not seen by fundus examination has become possible [178].

Because of eye motion, a minimal condition for convenient clinical use is to acquire an image of 200×200 pixels in a few seconds. This condition has been met in a device developed at MIT, now made commercially by Humphrey-Zeiss, using rather standard techniques in fiber optics. This technique is also promising for embryos in vivo, for which 2D imaging of a $3 \times 3 \times 3$ mm^3 volume with 15×30 μm spatial resolution has been demonstrated [179]. Limited success has been obtained in more strongly scattering tissues such as epithelia and arterial walls [180], using an optical-coherence microscope (OCM). This is a hybrid of a confocal microscope and a coherence-time tomograph, thereby combining geometric and temporal discrimination to reject out-of-focus information more efficiently. Images have been obtained down to a depth of 1–2 mm, though with poor spatial resolution.

Pushing techniques of in vivo microscopy to their limits seems clinically important. Over the last five years, experimental progress has been rather empirical: "We learned to focus and to process images", as once put by J. Izatt. Some theoretical work has been published, particularly to understand what exactly determines maximum penetration [176, 177, 181–184]. Penetration has been predicted to increase by a factor of two when coherent detection is used. However, these papers contain various restrictive assumptions, so that further consideration will be needed. A better understanding of the way in which incident light with known coherence properties interferes with a reference beam (heterodyne detection) after being scattered several times is desired. This problem is close to some aspects of coherent backscattering [15–17] and of coherent lidars (Chap. 5).

2.1.2 Oximetry

In a different area, a very useful application of light propagation in tissue is the pulse oximeter, which has now become widespread in hospitals. Using optical fibers, light is transmitted through the fingertip. As the absorption curves of oxygenated and deoxygenated hemoglobin differ significantly, a measurement with two IR wavelengths gives access – in vivo – to the relative concentration of the two, i.e. to hemoglobin oxygenation, and hence allows respiratory monitoring. Unfortunately, a severe limitation of standard oximeters is that they are only qualitative, and hence unacceptable in emergency situations or neonatology, for instance. The measured light attenuation is determined by the product of the absorption coefficient and the path length. The latter is not well defined in a finger, because it is affected by individual and temporal variations of scattering properties, blood content and geometry. As a result, accurate calibration of standard oximeters is impossible.

Some progress towards more quantitative measurements has been made with devices that allow separate quantitative determination of absorption and scattering coefficients. This is possible with light modulated at video frequencies with phase-and-amplitude-sensitive detection [185]. However, one useful property of the pulse oximeter [186], that it measures only the component at the heart frequency and thus provides information on arterial blood, is lost in the devices now proposed. A careful study, using our present understanding of light propagation in tissues, should allow the upgrade and optimization of present devices, and should eventually lead to significant clinical progress.

Recent progress in oximetry has included going from a volume-averaged measurement to a more local one. A typical example is the determination of muscle oxygenation, in which it is necessary to avoid perturbations due to blood in skin and superficial vessels. Such a measurement would be very useful for the study of pathologies. This diagnostic seems relatively simple – very-low-resolution imaging in an approximately stratified medium – and devices that have been proposed to minimize the influence of superficial regions on the measurement should help [187]. Such a regional, quantitative imaging device seems feasible in view of our knowledge of radiative transfer, yet it still requires serious work [188].

2.1.3 Fluorescence and Tumors

A potential application of optics concerns the early detection of superficial tumors. This is important because over 50 % of cancer diseases start in epithelia, i.e. in the first 100–200 µm below the surface. In situ detection at early stages, when the tumor is still confined to the epithelium, is of major importance. These stages are often difficult to discriminate from benign situations by simple visual inspection, so that partly random biopsies are needed. The fluorescence of tumoral cells has long been known to be generally different from that of normal cells, often much stronger. The problem is that both normal and tumoral cells have a broad variability in their fluorescence signal, so that simple fluorescence imaging is not going to be conclusive.

A useful new idea is that the fluorescence emission can be obscured by tumors located in the epithelium. The emission originates from below the epithelium and the thickness of the tumor exceeds that of the epithelium, thereby producing increased reabsorption. This is the case for the emission due to collagen, which is absent in epithelia. Again, this property alone is not specific enough, because epithelia thickening may also occur in benign situations. Recent progress in the case of the bladder seems to have been obtained with UV excitation at about 300 nm, presumably by a combination of emissions from inside and below the epithelia [189]. However, the features observed in the emission spectrum here are far from fully understood. Better understanding of the fluorescence and radiative transfer effects is needed, to determine the nature and origin of the various fluorescent emissions and absorption effects.

The use of drugs to produce exogenous fluorescence is now being actively studied, allowing imaging of the tumoral regions if a higher drug uptake can be obtained there. This method has the advantage of an increased emission intensity compared to autofluorescence, but it is accompanied by some clinical complications and side effects [190]. Another approach, two-photon fluorescence with infrared excitation, still has to be evaluated in detail [191].

A number of practical difficulties exist that have slowed down progress in tumor detection via fluorescence. Firstly, there is a large number of parameters. Secondly, we have only limited knowledge about fluorescence and radiative transfer in vivo. Thirdly, there is a large variability. Finally, early cancer stages are not readily available in vivo, because of the limited time for the experiments (done with a standard examination) and the difficulty with visual detection. Further progress, which is definitely possible and clinically of major importance, requires an increased cooperative effort: in vitro fluorescence studies on cell structures and on tissues, in vivo pointlike spectral studies in which the excitation and emission wavelengths are varied, in vivo tests of prototype fluorescence imaging systems and in vivo determinations of thicknesses and optical properties of epithelia and underlying tissues, all of this for various organs, and also a greater experience of doctors about what cellular stages should be distinguished.

2.1.4 Combining Ultrasonic and Light Waves

Preliminary work has established that when ultrasonic waves are focused in a scattering medium illuminated by light, modulation at the ultrasonic frequency (typically 1 MHz) can be detected in the transmitted light [102]. The modulated component seems to originate from the ultrasonic focal region, and to scale with both the local ultrasonic and the light intensity [192–194], Scanning the ultrasonic focus produces a photoacoustic image which should retain the submillimeter spatial resolution of the ultrasound and should even be able to locate objects without mechanical contrast, which makes them invisible in standard ultrasonic imaging. More work is needed to clarify the physical effects involved, to optimize the imaging [195] and to evaluate the clinical use of this technique.

To end this short overview, we note that other, miscellaneous applications of noninvasive optical techniques are also under study: blood velocimetry in the eye and skin, using either the Doppler effect or temporal correlations; measurements of the glucose concentration in the blood; and characterization of portwine stains before and during laser treatment.

2.2 Optical Imaging of Thick Organs

X-ray mammography, although widely used, has severe limitations. Its interpretation requires considerable training. X-ray screening is capable of detecting tumors as small as 1 cm [196], compared to the average actual size of detected tumors of 1.5 cm [197]. However, X-ray screening produces at least 20% of false negatives (i.e. undetected tumors) and about 90% of false positives [196]. Methods able to reliably detect 0.5 cm or even 1 cm tumors would be most welcome. Can IR light do this? While IR can be detected through the breast, scattering produces highly blurred transmission images: the point spread function (PSF) of an inhomogeneity located deep in the breast is several centimeters wide. As a result, simple IR transmission imaging is insufficient.

Over the last 15 years, efforts to reduce the PSF width, mainly using picosecond pulsed lasers, have given only limited success. Although a reduction factor of 1.5–2 has been obtained by various methods, further improvement involves a drastic reduction of the detected signal. During the same period, brute-force numerical simulations of the inverse problem (the numerical reconstruction of the absorption and/or scattering distribution inside the organ from information obtained using a set of external light sources and detectors) have led to huge computation times. More recently, faster (but approximate) inversion techniques have been proposed. At the same time, it has been recognized that tumor detection is not only a matter of resolution, but also one of contrast, and that both the absorption and the scattering coefficients may be affected in benign and malignant tissue modifications. While measuring both coefficients may well provide more information and specificity, it also makes the problem more difficult. As we will see, more parameters generally imply a decreased accuracy for each separately, as well as a larger minimum size for which the inhomogeneity can be characterized.

New prototypes of optical mammographers have recently been built by several companies and are now being tested in hospitals [198–200]. This demonstrates the continuing clinical interest in optical mammography. Other imaging prototypes have been built for neonatology [201] to detect brain hematoma [202] and to measure brain activity [203, 204].

2.2.1 Forward Problem

Since the dimensions of interest are much larger than the transport mean free path, the diffusion equation can be used. In a piecewise uniform medium, this equation has the form of a modified Helmholtz equation. In our field it is usually written as

$$\frac{1}{c}\frac{\partial N}{\partial t} - D\nabla N + \mu_{\mathrm{a}}N = S. \tag{2.1}$$

N is the "photon density" and $D \equiv [3(\mu_{\mathrm{a}} + \mu'_{\mathrm{s}})]^{-1}$ is usually called the diffusion constant, although it has the dimensions of length, not length^2/time as is customary in other fields. S denotes a source term. The first term in (2.1) disappears for a stationary source, and can be merged into the third term for a frequency-modulated source (in which case $\mu_{\mathrm{a}} \to \mu_{\mathrm{a}} + i\omega$). A collimated incident laser beam can be approximated by a point source, and the boundary conditions by image sources [172]. Similar equations are common in heat transfer (with $\mu_{\mathrm{a}} = 0$) and electrostatics (where $D \to -D$) [205]. Several recent advances in solving (2.1) came from its applicability in other fields.

A necessary condition for a fast numerical solution of the inverse problem is the existence of a fast solution of the forward problem, that is, the computation of the detector signals for given absorption and scattering distributions in the medium. The simplest possible calculation is given by the Born approximation. Average absorption and scattering properties $(\mu_{\mathrm{a}0}, \mu'_{\mathrm{s}0})$ are defined for the medium to be imaged, and a voxel with parameters $(\mu_{\mathrm{a}0}+\Delta\mu_{\mathrm{a}}, \mu'_{\mathrm{s}0}+\Delta\mu'_{\mathrm{s}})$ is treated as a perturbation. This approximation is certainly accurate if the following conditions are met: only one voxel differs from the background, and its volume V and its contrasts $\Delta\mu_{\mathrm{a}}/\mu_{\mathrm{a}0}$ and $\Delta\mu'_{\mathrm{s}}/\mu'_{\mathrm{s}0}$ are very small. Obviously, however, these conditions are not necessarily met in tissues. So we want to know whether they can be relaxed.

Consider first the effect of a finite contrast. For a small inhomogeneity (one single voxel) in a homogeneous medium we can demonstrate, using the so-called Lorentz factor known in electrostatics [205], that the Born approximation applies for large contrasts $\Delta\mu_{\mathrm{a}}/\mu_{\mathrm{a}0}$ and $\Delta\mu'_{\mathrm{s}}/\mu'_{\mathrm{s}0}$, provided that the volume V is sufficiently small. The correction factor becomes significant if the scattering contrast exceeds unity.

Next, consider the effect of a finite volume. We have compared an exact solution available for a spherical, uniform inhomogeneity (which corresponds to the Helmholtz equation) to the multivoxel solution obtained by decomposing the sphere into small voxels and adding the Born solution of each voxel. This approximation resembles somewhat the Rayleigh–Gans approximation known in optics [206]. We found good agreement for the image width, and concluded that the error in the amplitude can be reduced using a simple quasi-linear correction [207]. This means that it is possible to avoid nonlinear, time-consuming models.

2.2.2 Inverse Problem

As to the inverse problem, *detection* and *characterization* should be distinguished [207]. We first consider a purely absorbing inhomogeneity, which we want to image with continuous-wave (CW) illumination. The inhomogeneity can be detected if it significantly modifies the measured signals compared to the noise level. The smaller the inhomogeneity, the smaller its influence on the detector signals. For a given geometry and given contrast conditions this leads to a minimum size of the inhomogeneity, called the detection threshold.

Characterizing this inhomogeneity means finding its volume and absorption coefficient from the amplitude and width of the image. From the solution of the forward problem we know that the amplitude is approximately proportional to the volume and contrast, whereas the width is a weak function of volume. For a given accuracy of the measurements, one can estimate from the forward problem the error bars of the deduced volume and contrast. A minimum size of the inhomogeneity is required in order to keep these error bars reasonably small, say 30 %. This characterization threshold is significantly larger than the detection threshold.

When several properties of the background and inhomogeneity differ at the same time, the threshold picture is more complex. Again, we assume that the accuracy of the measurements is known; it is typically about 0.1 % in amplitude and 0.1° in phase. A systematic method is needed to evaluate the accuracy of the quantities to be reconstructed, namely the volume, absorption and scattering coefficients; we require in particular the minimum size for which an accuracy better than around 30 % can be obtained for all parameters, i.e. a global characterization threshold.

Such a threshold is provided by the Bayesian approach [209]. The measurements yield a data set $\mathbf{M}$, to be considered as a Gaussian random vector with average $\mathbf{M}_0$ and covariance matrix $\mathbf{B}$,

$$B_{ij} \equiv \langle \delta M_i \delta M_j \rangle , \tag{2.2}$$

where $\delta M_i \equiv M_i - M_{i0}$ is the fluctuation of M_i. This method provides, for the parameter set of unknowns $\mathbf{P} \equiv (V, \mu_{\mathrm{a}}, \mu_{\mathrm{s}}')$, a mean value $\mathbf{P}_0$ as well as a covariance matrix $\mathbf{C}$. In addition to the most probable values, standard deviations are obtained for all unknowns P_i, as well as for any combination of them.

When continuous illumination is used, our work shows that inhomogeneities that both absorb *and* scatter are much more difficult to characterize than those that only absorb or only scatter. The diameter of a spherical inhomogeneity must be larger than 2 cm in the first case, compared to 1.2 cm in the latter two cases. A size of 2 cm is unacceptably large for early cancer detection. On the other hand, if the illumination is modulated at 200 MHz, the minimum size reduces to 1.2 cm in all cases, which is marginally acceptable. The results can be put as follows: for a size of 1.2 cm, for which the

CW image provides information on two unknowns, adding the phase image provides information on three unknowns with comparable accuracy.

Fortunately, it is possible to say "something" about an inhomogeneity with a size less than the characterization threshold. We shall refer to this as partial characterization. We have considered the angle θ, related to the contrast ratio $\Delta\mu_s'/\Delta\mu_a$ and defined in the (μ_a, μ_s') plane as the angle (in the range 0 to π) between the μ_a-axis and the line BM, where B represents the background point (μ_{a0}, μ_{s0}') and M the inhomogeneity (μ_a, μ_s'). We found that θ remains accurate within an uncertainty of 20 degrees down to sizes of 1.2 cm for continuous illumination and to 0.8 cm in the case of frequency modulation. Consequently, some information on the nature of the inhomogeneity remains available below the characterization threshold. By "nature" we mean the distinction between scattering and absorption and between positive and negative contrasts. This can be understood from the fact that the image width depends more on the nature of the inhomogeneity than on its size, at least when its radius is smaller than 2 cm.

As has been noted earlier, the information actually available on the contrasts in human organs [210, 211] is insufficient. Therefore, any prediction of the clinical value of the parameter θ is simply impossible. Absorption seems to increase in both benign and malignant deseases, whereas scarce and contradictory results have been reported for scattering. The Bayesian approach that we have used should be applied in many other situations. A typical example is the joint determination of several parameters from a set of experiments, such as μ_a and μ_s' from spatially or temporally resolved reflectance curves, for which published estimates of the error bars are probably too optimistic.

In conclusion, among the optical diagnostic methods contemplated, microscopy, oximetry and fluorescence have been proved useful in clinical medicine. The structural and functional imaging of large organs is now developing along these lines, with several steps forward to be expected in the near future.

3. Effective-Medium
and Coherent-Potential Approximation

Kurt Busch and Costas Soukoulis

As is evident from previous contributions in this volume, microscopic multiple-scattering calculations are even more cumbersome for classical than they are for electron waves. This has generated a considerable amount of interest in developing an effective-medium description for classical wave propagation in strongly scattering disordered media. However, early approaches based on the analogy between classical and electron waves have had only limited success [54, 134, 212–214]. In particular, the problem of efficient and reliable calculation schemes for the energy transport velocity proved to be a hard nut to crack.

In the first part of this chapter (Sect. 3.1) we outline the reasoning that led us to the formulation of a very successful, new effective-medium theory which is based on the principle of energy homogeneity. The main results and predictions of this new effective-medium theory are summarized in the remainder of the chapter. At this point, we want to stress that this theory can be applied to a great variety of problems in the field of classical wave propagation.

3.1 The New Effective-Medium Theory

Our composite medium is assumed to consist of randomly placed lossless spheres, with diameter $d = 2R$ and dielectric constant ϵ_1, embedded within a lossless host material with dielectric constant ϵ_2. The random medium is characterized also by f, the volume fraction occupied by the spheres.

The basic idea of any effective-medium theory of disordered systems is to focus on one particular scatterer and to replace the surrounding random medium by an effective homogeneous medium. The effective medium is determined self-consistently by taking into account the fact that any other scatterer could have been chosen. This procedure manifests the homogeneity of the random medium on average.

However, the position of a sphere in the medium is completely random, with the exception that the spheres cannot overlap. This implies that the distribution $P(R)$ of spacings between neighboring spheres is sharply peaked at a distance $R_c > R$. If we approximate this distribution by a δ-function, i.e., $P(R) \propto \delta(R_c - R)$, and take into account the on-average isotropy of the random medium, we may consider a coated sphere as the basic scattering unit. The radius R_c of the coated sphere is $R_c = R/f^{1/3}$. The dielectric constants of the core and the coating are ϵ_1 and ϵ_2 respectively. Using a coated sphere as the basic scattering unit also incorporates some of the multiple-scattering effects at different centers.

The use of a coated sphere as the basic scattering unit also implies that the homogeneity of the energy density is not fulfilled any more. This fact has not been taken into account in approaches that exploit the analogy between classical and electron wave propagation [54, 134, 212–214], and as a consequence, lead to unphysical results for the transport velocity. Therefore, in the new effective-medium theory [55, 215, 216] we explicitly chose the averaged energy density homogeneity as the criterion for determining the effective-medium. Since we are exclusively considering lossless dielectrics, the effective-medium dielectric constant $\bar{\epsilon}$ has to be real owing to energy conservation. This is in contrast to the conventional coherent potential approximation (CPA) approaches [54, 134, 212–214] and forces us to proceed in two steps. Firstly, we determine, for every frequency ω, the real effective dielectric constant $\bar{\epsilon}$ by demanding the energy density to be homogeneous on scales larger than the basic scattering unit (coated sphere). Then, in a second step, the physical quantities are calculated from the (now nonvanishing) scattering cross-sections. In this theory all multiple-scattering effects are contained in the effective dielectric constant and, thus, we may consider the random medium as consisting of independent scattering units, i.e., coated spheres, embedded in the effective-medium. Figure 3.1 schematically depicts the reduction of the disordered medium to a description of independent coated spheres embedded in the effective-medium.

The requirement that the energy content of a coated sphere, embedded in the effective-medium and being illuminated by a plane wave, should be the same as the energy stored by the plane wave in an equally sized volume of the effective-medium can be formulated quantitatively by the self-consistency equation

$$\int_0^{R_c} \mathrm{d}^2 r \; \rho_E^{(1)}(\mathbf{r}) = \int_0^{R_c} \mathrm{d}^2 r \; \rho_E^{(2)}(\mathbf{r}) \,, \tag{3.1}$$

where $\rho_E^{(1)}(\mathbf{r})$ and $\rho_E^{(2)}(\mathbf{r})$ are the energy densities for a coated sphere and a plane wave respectively. Clearly, this very general principle can be applied to any kind of classical wave propagation, including elastic waves [53].

The specific forms of the scattered fields inside the coating and the core are given in [215] for the three-dimensional case and in [216] for two spatial dimensions. At this point it suffices to note that while the right-hand side of (3.1) is an obvious function of the effective-medium dielectric constant $\bar{\epsilon}$, the left hand side, through its complicated dependence on the fields' scattering coefficients, is a nontrivial function of the effective-medium dielectric constant $\bar{\epsilon}$, too. Consequently, (3.1) and the respective expressions for the fields determine the (real) dielectric constant $\bar{\epsilon}$ of the effective-medium for every

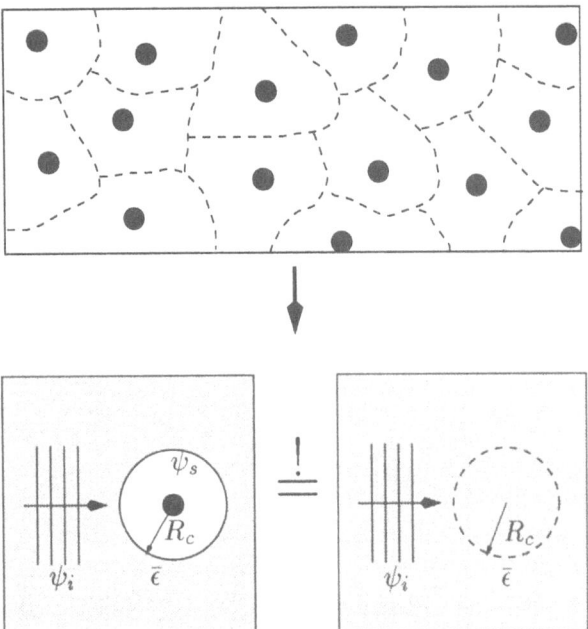

Fig. 3.1. In a random medium composed of spheres with dielectric constant ϵ_1 immersed in a host medium with dielectric constant ϵ_2, the basic scattering unit may be, on average, regarded as a coated sphere, as represented by the *dashed lines*. To calculate the effective dielectric constant $\bar{\epsilon}$, a coated sphere of radius $R_c = R/f^{1/3}$ is embedded in a uniform medium. The self-consistent condition for the determination of $\bar{\epsilon}$ is that the energy of a coated sphere is equal to the energy of a sphere with radius R_c and dielectric constant $\bar{\epsilon}$ [reprinted from Busch and Soukoulis, Phys. Rev. Lett. **75**, 3442 (1995), with permission from the American Physical Society]

frequency. As mentioned above, the energy transport velocity, the renormalized wave vector $\bar{k}$ and the scattering mean free path ℓ can now be calculated from the resulting self-energy of independent coated spheres embedded in the effective-medium [55, 215, 216]. In addition, we approximate the transport mean free path ℓ^* by the scattering mean free path ℓ, i.e. $\ell \approx \ell^*$. Then, the three-dimensional diffusion constant D is given by $D = v_E \ell^*/3$. This approximation is supported by the fact that, as a mean-field theory, the new effective-medium theory is unable to make detailed predictions close to the Anderson transition, where the distinction between the scattering and transport mean free paths would become important. In addition, previous studies of the transport and scattering mean free paths [134] have obtained results consistent with this approximation.

3.2 Results for Long-Wavelength Limit

In the long-wavelength limit, we may define a frequency-independent, long-wavelength dielectric constant ϵ_∞ according to

$$\epsilon_\infty = \lim_{\omega \to 0} \left(\frac{c}{v_E(\omega)} \right)^2 . \tag{3.2}$$

It is well known [217] that for scalar classical waves the correct result for ϵ_∞ is given by the volume-averaged dielectric constant, whereas in the case of vector classical waves it is the Maxwell–Garnett theory which gives the right answer. An analytical calculation [215, 216] of ϵ_∞ within the new effective-medium theory, using (3.2), proceeds straightforwardly by computing $\bar{\epsilon}$ for $\omega \to 0$ from (3.1), using a Taylor expansion of all quantities involved to extract the leading order in ω. Indeed, in the case of scalar classical waves we obtain for the long-wavelength dielectric constant the volume average of ϵ_1 and ϵ_2, i.e.

$$\epsilon_\infty \equiv \bar{\epsilon} \equiv f\epsilon_1 + (1 - f)\epsilon_2. \tag{3.3}$$

This result originates from the fact that for scalar waves s-wave scattering dominates in the long-wavelength limit.

In the case of a vector polarization, however, s-wave scattering is absent and a careful analysis [215, 216] of the dominant p-wave scattering for long wavelengths leads to the Maxwell–Garnett result, i.e.

$$\epsilon_\infty = \bar{\epsilon} = \epsilon_2 \left(1 + \frac{d f \alpha}{1 - f \alpha} \right) , \tag{3.4}$$

where $d = 2, 3$ stands for the dimensionality of the system, and the depolarization factor α of a sphere or cylinder is given by $\alpha = (\epsilon_1 - \epsilon_2) / [\epsilon_1 + (d-1)\epsilon_2]$.

3.3 Results for Finite Frequencies

For finite frequencies, of course, no analytical solution of (3.1) is possible. Fortunately, it turns out that (3.1) is numerically easy to deal with.

Figure 3.2 shows a comparison between the diffusion coefficient $D = v_E \ell^* / 3$ obtained within the new effective-medium theory and experimental values obtained by Genack et al. [218]. Without adjustable parameters, excellent agreement is obtained between experiment and theory. More detailed results on the energy transport velocity and mean free paths within the new theory can be found in [55, 215, 216].

3.4 Study of the Localization Parameter

The product $\bar{k}\ell^*$, where $\bar{k}$ is the renormalized wave vector and ℓ^* the transport mean free path, is a measure of the strength of the multiple-scattering

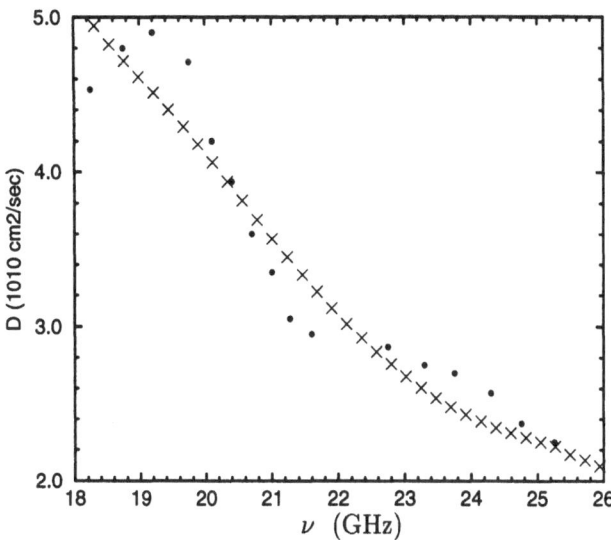

Fig. 3.2. Frequency dependence of the diffusion constant for a sample of 13 mm polystyrene spheres with index of refraction 1.59 and filling ratio $f = 0.59$. The *filled circles* correspond to the experimental values, and the *crosses* are the results of the new effective-medium theory [55] [reprinted from Busch and Soukoulis, Phys. Rev. B **54**, 893 (1996), with permission from the American Physical Society]

effects. Here, we wish to recall that within the effective-medium theory we have assumed that $\ell^* \approx \ell$. For values of $\bar{k}\ell^* \simeq 1$ coherent backscattering significantly renormalizes the diffusion coefficient and may ultimately lead to a change in the nature of wave functions from extended to localized. This phenomenon, called Anderson localization [3], is a generic wave property and it is still an open problem whether classical waves can be localized. There exist various theories which provide localization criteria for waves: if the value of $\bar{k}\ell^*$ falls below a certain value, localization is achieved. Probably one of the most accurate among these is the potential-well analogy [219], which gives the critical value for $\bar{k}\ell^*$ as 0.844.

Clearly, in a mean-field theory like the new effective-medium theory no quantitative statements as to when a wave system crosses from extended to localized can be made. However, the value of the localization parameter $\bar{k}\ell^*$ can still be evaluated and, as a function of the system parameters, it may exhibit certain trends towards the parameter values that are optimal for localization. In this spirit, we have performed a systematic study of the localization parameter $\bar{k}\ell^*$ as a function of the dielectric contrast ϵ_1/ϵ_2 and filling fraction f for electromagnetic waves, for the direct ($\epsilon_1 > \epsilon_2$) as well as for the inverse ($\epsilon_2 > \epsilon_1$) structure. We determined for every combination of parameter values the minimum of $\bar{k}\ell^*$ as a function of frequency. In this way we were able to obtain contours of constant $\bar{k}\ell^*$ as a function of the dielectric contrast and filling fraction. The results for two- and three-dimensional systems may be found in [216]. For two-dimensional systems, we find that in the direct structure scalar waves are easier to localize than vector waves, whereas the opposite is true in the inverse structure. For the direct structure the optimal filling ratio for both types of wave is around $f \approx 0.25$, whereas for the inverse structure the scalar wave has its optimal filling ratio at $f \approx 0.8$, while the vector wave has its optimum around $f \approx 0.6$. Furthermore, the values of $\bar{k}\ell^*$ achieved within the same parameter range are much lower for the inverse structure than they are for the direct structure. This result has already been obtained on the basis of numerical simulations [220]. Similarly, in three spatial dimensions, we find that localization for electromagnetic waves may be easier to achieve for the inverse structure, the optimal filling factor f is approximately 0.45 for the direct and 0.7 for the inverse structure.

3.5 Discussion

In summary, the new effective-medium theory allows one to reliably calculate transport properties of disordered classical wave systems. It can be applied to a wide variety of wave propagation problems. In the long-wavelength limit, well-known results are recovered. Without adjustable parameters, excellent qualitative as well as quantitative agreement with experiment is obtained. A study of the localization parameter $\bar{k}\ell^*$ predicts the optimal parameter ranges for classical localization, depending on the type of wave (scalar or vector) and type of structure (direct or inverse configuration). Generally, localization is favored in the inverse structure, since the localization parameter $\bar{k}\ell^*$ takes on lower values in the inverse structure.

4. Acoustic and Ultrasonic Propagation

4.1 Ultrasonic Wave Propagation in Random Media

Arnaud Derode, Mathias Fink, Philippe Roux and Jean-Louis Thomas

Electrons, photons and ultrasound are described by a wave equation. Therefore many results derived in solid-state physics or optics must hold for acoustics as well. The aim of this contribution is to present the experimental advantages of ultrasonic waves compared to other types of wave [221]. The relative slowness of acoustic waves – commonly between 10^2 and 10^4 m/s, and much smaller than the 3×10^8 m/s for light – as well as their relatively low frequency (ranging from about 1 kHz for audible sounds to 10 MHz for ultrasound, compared to 10^{14} Hz in the case of visible light) has deep consequences for the perception of these phenomena. Indeed, unlike light waves, acoustic phenomena are slow enough for the detectors to record their real-time variations, whereas optical detectors are only intensity-sensitive. Reciprocally, for transmitters, controllable wide-band ultrasonic sources (piezocomposite arrays) exist that can create any waveform, whereas optical wave packets always depend on how light emission is performed by atoms. In a way, acoustics has the same capabilities that optics would offer if a 10^{15} Hz oscilloscope and a perfectly controllable light source existed. We present three applications below.

4.1.1 Coherence Measurement and Aberration Compensation

We use wide-band N-element programmable arrays, each element of which can transmit or receive any signal, and record the fluctuations of a pressure field $f(x,t)$.

Direct measurement of the degree of spatial coherence of the wavefield, without using interferometric processes, is possible with these arrays [222, 223]. If we consider two detectors A and B, the degree of spatial coherence can be quantified simply by the correlation coefficient

$$\rho(A,B) \equiv \frac{\int_T \mathrm{d}t f(A,t) f(B,t)}{\sqrt{\int_T \mathrm{d}t f^2(A,t) \ \int_T \mathrm{d}t f^2(B,t)}}, \tag{4.1}$$

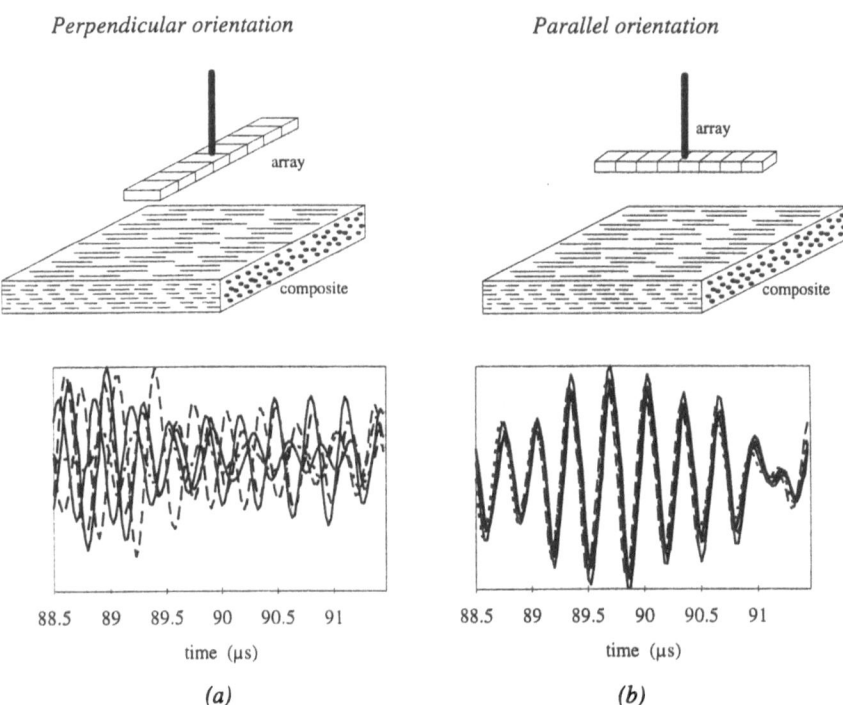

Fig. 4.1. Experimental results for a unidirectional composite: backscattered signal received by elements 1, 4, 8, 12 and 16 in perpendicular (a) and parallel (b) orientation. When the array is parallel to the fibers the backscattered signals are almost identical, as if the incident wave felt the coherent aspects of the scattering medium [reprinted from Derode and Fink, *Spatial Coherence of Ultrasonic Speckle in Composites*, IEEE Tr. on Ultras. Ferr. & Freq. Control, Vol. 40, No. 6, © 1993 IEEE]

with T some time window of interest. The variation of this correlation coefficient as a function of the distance AB between the receiving elements depends on the statistical properties of the medium.

A simple and striking example is given in Fig. 4.1. A 16-element focused array sends a short ultrasonic pulse into a heterogeneous medium. The sample is a unidirectional composite, made of a regular stacking of carbon fibers embedded in an epoxy resin. The fibers, whose diameter is much smaller than the average wavelength, scatter the incoming wave in all directions. Consequently the array records 16 backscattered signals. From these signals, we can calculate the degree of coherence of the backscattered field as a function of the distance between two elements. It strongly depends on the orientation of the fibers: according to whether the array is parallel or perpendicular to the fibers, the backscattered field is more or less coherent. This is obvious when looking at the backscattered signals.

Piezoelectric arrays are ideal tools to perform real-time adaptive processing. For instance, the presence of aberrations is a major cause of image degradation (in medical imaging, this can be due to subcutaneous fat layers). These aberrations induce unknown time delays that distort the ultrasonic beam (Fig. 4.2). A technique to correct for these effects uses a measurement of the degree of spatial coherence with a transducer array. Most biological tissues act as random, single-scattering media for ultrasound; within the single-scattering approximation, one can easily calculate the degree of spatial coherence of backscattered waves: the coherence length is inversely proportional to the size of the scattering volume. More precisely, with a focused linear array of length a, the correlation coefficient between the signals received at A and at B decreases linearly as the distance AB increases, and is zero when $AB > a$. This triangular profile can be observed experimentally (Fig. 4.2a).

In the presence of an aberrating layer (Fig. 4.2b), the transmitted wavefronts are distorted, the ultrasonic beam is no longer correctly focused, the focal spot is wider and the backscattered correlation curve becomes narrower. We try to compensate for the effects of the aberrating layer by calculating the correlation coefficient between two neighboring elements with an additional time delay τ. The value of τ that maximizes the correlation coefficient is the time delay induced by the aberrator. We repeat this procedure for all elements of the array to obtain a set of time delays. The aberration effects can be compensated by programming independently the elements of the array: as is shown in Fig. 4.2c, the ultrasonic beam is now well focused, and the backscattered correlation curve recovers its triangular shape.

This technique is similar to those used in astronomy to compensate for the effects of the atmosphere, such as deformable mirrors or arrays of mirrors, although it needs no 'reference point' (a bright star). The measurement of the coherence properties of the speckle noise itself is sufficient.

4.1.2 Coherent Backscattering with Acoustic Waves

The coherent backscattering cone is a spectacular manifestation of persistent coherence effects despite high-order multiple scattering. It appears as a peak in the scattered intensity: in the backscattered direction, the intensity is (roughly) twice as large as in other directions. This interference effect is due to the only invariant property in a static disordered system: time-reversal symmetry. All possible multiple scattering paths add up incoherently, except near the backscattering direction, where the incoming wave and its conjugate are necessarily in phase, and give rise to a constructive interference. In recent years, this phenomenon has been observed with visible light as explained in detail in Sect. 1.4.1. The effect has been reported for acoustic waves as well [32].

We have observed coherent backscattering with a 128-element piezoelectric array [224]. The time-resolved data were integrated over time to get

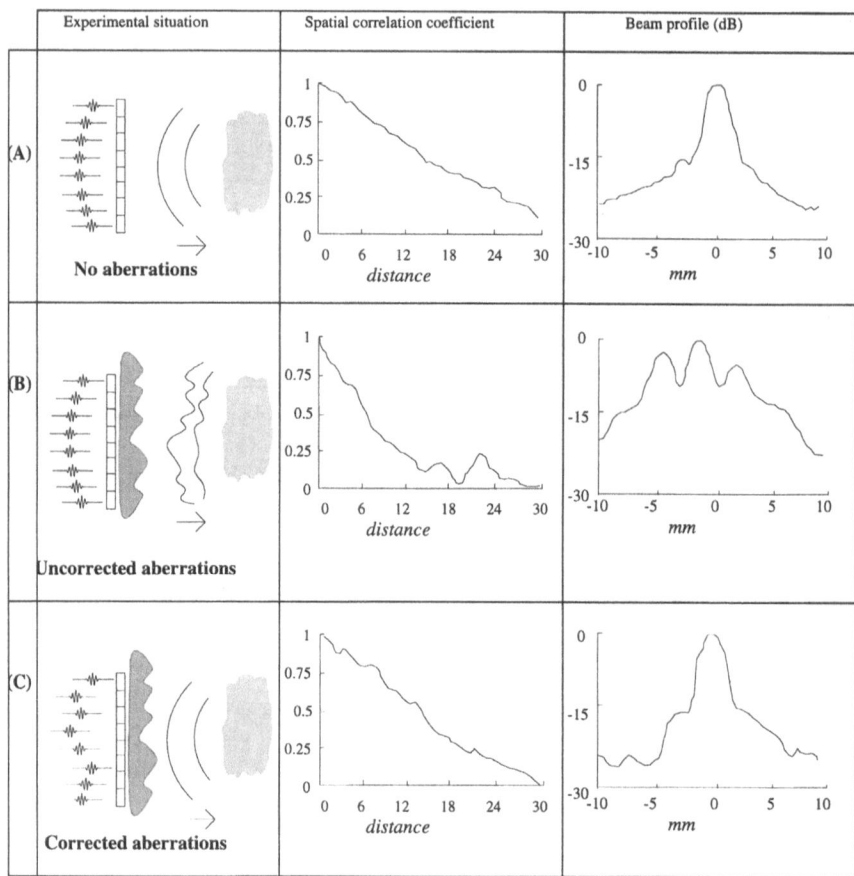

Fig. 4.2. Adaptive aberration compensation. Without an aberrating layer (**a**), the beam is well focused and the backscattered correlation curve has a triangular shape. In the presence of an aberrating layer (**b**) the beam is distorted and no longer correctly focused, and the backscattered correlation curve is narrower. By finding the appropriate time delays (**c**), one can compensate for aberrations and focus the beam properly

the stationary coherent backscattering and thus the acoustic equivalent of Fig. 1.4. Note that "backscattering direction" in optics translates here to "source location". Whereas the optical experiment is an experiment with plane waves and is thus carried out in Fourier space, the present acoustic experiment was performed in real space. One of the elements (No. 51) sent a pulse into a scattering medium, and all elements recorded the backscattered signals. The sample was a set of roughly 2000 parallel steel rods with diameter 0.8 mm (the average wavelength was $\lambda = 1$ mm). Ensemble averaging was obtained by shifting the array and acquiring new data. Fig. 4.3 presents a backscattering intensity curve obtained from 400 data sets (presumably

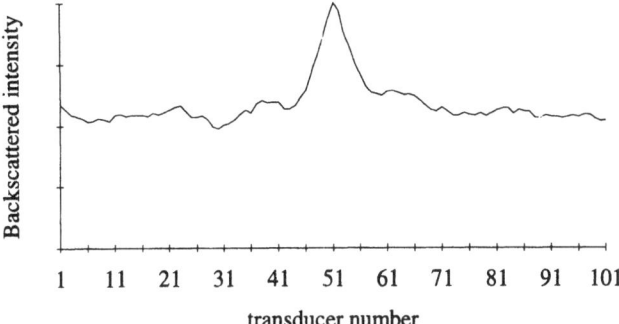

Fig. 4.3. Stationary coherent backscattering with acoustic waves. The average backscattered intensity at the transmitting transducer (No. 51) is roughly twice as large as the background level. The angular width of the peak is 15.6 mrad (for more details see [224])

independent). The backscattering peak is very visible; its FWHM is $\theta = 15.6$ mrad, which leads, according to (1.17), to an estimate for the transport mean free path $\ell^* = 10.2$ mm, in very good agreement with other measurements.

4.1.3 Time Reversal of Wave Propagation

Ultrasonic arrays can be used as time-reversal mirrors (TRMs) [225, 226]. Basically, a TRM is a device that records a waveform $f(t)$, time-reverses it and retransmits $f(-t)$ into the medium. An ultrasonic TRM is an array of N wide-band piezoelectric transducers that performs a fine spatial sampling of the acoustic field. The N signals are recorded, placed into a memory, time-reversed and retransmitted through the medium by the same transducer array. The medium we investigated exhibits very-high-order multiple scattering; it is a random set of 2000 parallel steel rods, the average distance between two rods being 2.6 mm, and their diameter 0.8 mm. The estimated transport mean free path is 10 mm.

The general principle of the experiments is presented in Fig. 4.4. An ultrasonic source S transmits a short pulse that propagates through the rods. The TRM receives a set of signals and records them into electronic memories. These signals are time-reversed and then retransmitted by the mirror, thus creating an ultrasonic wave that propagates through the same sample. Finally, the pressure field is measured at the source S.

The mirror was a linear array of 64 transducers with central frequency 3.5 MHz, corresponding to an average wavelength $\lambda \approx 0.43$ mm in water, bandpass 50 % and pitch 417 μm. In each channel, the signals were sampled at 20 MHz, digitized at 8 bits and recorded. The source S was a single transducer (size 387 μm, central frequency 3.5 MHz), and the sample was located between the source S and the mirror. After the time-reversal process, the same transducer S recorded the pressure field as a function of time, but it

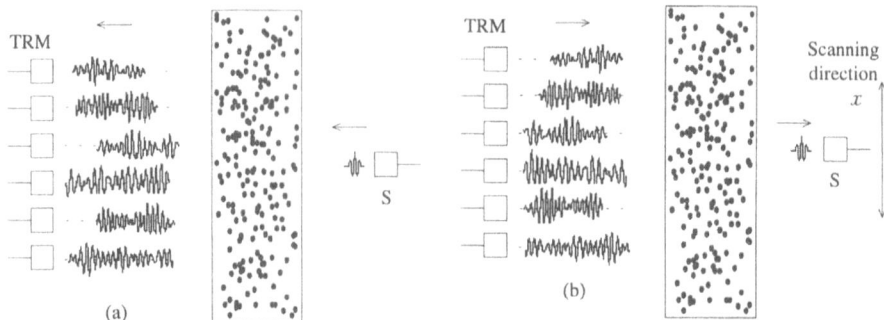

Fig. 4.4. Experimental setup for ultrasonic time reversal. (a) First step: S sends a pulse through the sample, and the transmitted wave is recorded by the TRM. (b) Second step: after the scattered signals have been time-reversed, they are retransmitted by the TRM, and S records the reconstructed pressure field. The operation can be repeated as S is translated along the x-axis in order to scan the pressure field

could also be translated along the direction of the array (x-axis (vertical) in Fig. 4.4) in order to scan the pressure field. So S measures a two-dimensional signal $s(x, t)$. Spatial resolution is measured by a directivity pattern $d(x)$: the maximum value for each position was detected, which provided the directivity pattern $d(x) = \max_t s(x, t)$.

Using short ultrasonic signals, it was possible to separate the "ballistic" part of the transmitted field, which arrived first at the array, from the multiply scattered parts. The transmission coefficient for the incoherent intensity $T(L)$ could be measured experimentally with wide-band transducers (central frequency 3.5 MHz) as a function of the thickness L. The slope of the curve on a logarithmic scale leads to an estimate of the transport mean free path $\ell^* = 7 \pm 0.5$ mm, according to the formula (1.14), known in optics and electronics.

In the following experiments, the thickness was fixed at $L = 45$ mm. This distance is greater than the mean free path, which indicates a multiple-scattering regime. This is indeed obvious when looking at the transmitted signals (Fig. 4.5a): after the coherent ballistic wavefront a very long signal appears; it can last over 200 μs after the arrival of the coherent wavefront, and corresponds to multiple paths with lengths of up to 300 mm inside the sample (i.e. 40 times the mean free path, or 120 times the average spacing between two rods).

The directivity pattern of the TRM was measured in water. In the absence of any scattering medium, the array receives a spherical wavefront coming from the source S. Then the time-reversed retransmitted wave converges towards the source. The signal received by S is represented in Fig. 4.5d, and the corresponding directivity pattern is plotted in Fig. 4.6; it exhibits symmetrical sidelobes as predicted by diffraction theory, and its width at

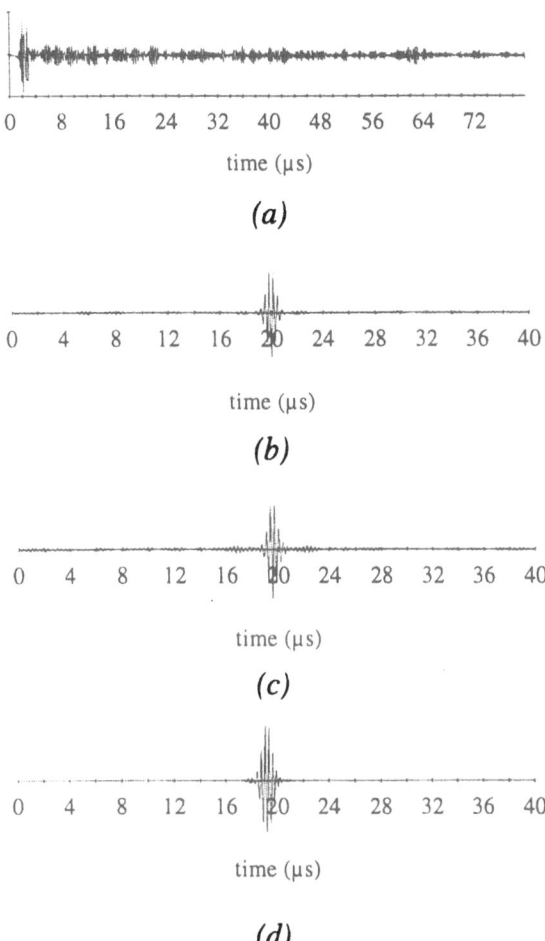

Fig. 4.5. (a) Waveform received through the rods at the TRM by element No. 32. (b) Waveform received by S after time-reversing the first 80 μs of the multiply scattered signals. (c) Waveform received by S after time-reversing an 80 μs time window beginning 40 μs after the ballistic front. (d) Waveform received by S after time-reversing the signals in the absence of any scattering medium. (The time origins have been chosen arbitrarily) [reprinted from Derode, Roux and Fink, Phys. Rev. Lett. **75**, 4206 (1995), with permission from the American Physical Society]

half maximum is $\Delta = 6.3$ mm. The theoretical value is $1.22\,\lambda z/a = 6.38$ mm, with $z = 330$ mm and $a = 64 \times 0.417$ mm $= 26.7$ mm.

If we place the rods between the point source and the mirror, the signals received at the mirror become much longer (Fig. 4.5a). The first 80 μs of the 64 signals were recorded, time-reversed and retransmitted (the playback resolution is the same as the digitization resolution, 8 bits). The resulting acoustic wave traversed the sample. An amazing compression was observed

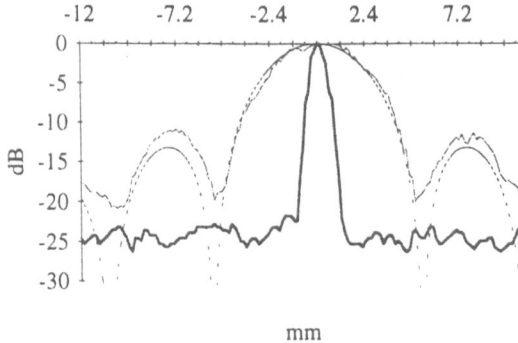

Fig. 4.6. Directivity patterns of the TRM through 2000 steel rods (*thick line*) and in water alone (*thin line*). The theoretical sinc function is plotted as *dashed lines* [reprinted from Derode, Roux and Fink, Phys. Rev. Lett. **75**, 4206 (1995), with permission from the American Physical Society]

(Fig. 4.5b); the received signal lasts about 1 μs, against more than 200 μs for the time-reversed multiply scattered signals. The same experiment was repeated with a different time window; this time, the TRM time-reversed and retransmitted 80 μs of the multiply scattered signal, beginning 40 μs after the ballistic front, and thus contained only multiply scattered contributions. As is shown in Fig. 4.5c, the initial pulse could still be reconstructed at the source S. In both cases, the time-reversed waves seem to have effectively gone backwards through the same paths, and found their way back to the source.

The final experimental result (Fig. 4.6) shows the influence of high-order scattering on the spatial resolution [226]. It was obtained using a set of 2400 steel rods (with an average spacing of 2.5 mm, the overall thickness of the sample being $L = 75$ mm) by reversing a 235 μs time window beginning 10 μs after the "ballistic" front. The resolution we observed was 1.05 mm, that is to say one-sixth of the theoretical limit for a homogenous medium (6.38 mm). There are no more sidelobes, but, rather, a uniform background; its level is roughly -24 dB, in comparison to -13.5 dB for the classical "sinc" function. This increase in the resolution can be explained by the large lengths of multiple-scattering paths involved. As a matter of fact, multiple-scattering widens the acoustic beam, thus creating a halo that is much larger than in the homogeneous case. According to the Ornstein–Zernicke theorem for spatial correlations [64], the resolution is inversely proportional to the halo surface area. A plausible explanation for the observed autofocusing is that when the time-reversed field is retransmitted by the array, the sample acts as a source with a larger angular diameter, which accounts for the enhanced resolution.

Thus, there is the paradox that multiple scattering that may be expected to degrade the directivity of a wave, as well as its temporal resolution, can in fact help to refocus this wave. An amazing time and space compression is observed, the accuracy of focusing being even better than the theoretical resolution for a diffraction-limited aperture!

4.2 Coherent Beam

Ping Sheng

Calculating the wave velocities of a resonantly scattering medium is a classical problem that has been recognized in the works of Sommerfeld [227] and Brillouin [228]. Recently, Page et al. [229] reported on work which resolved this problem in the case of ultrasonic wave propagation in samples consisting of packed glass beads immersed in water. This work consisted of both an experimental and a theoretical component.

Experimentally, the aim was to measure the group velocity and the phase velocity of the system. In order to observe the coherent component of the wave, thin, disc-shaped slabs of monodisperse glass beads were used. The glass beads were either 1 mm or 0.5 mm in diameter. Since the acoustic wavelength is comparable, the system is in the resonantly scattering regime, and the scattering mean free path is of the order of the bead size. The thickness of the samples ranged from 2 to 5 mm. Each sample was immersed in water, with transducers generating a plane-wave pulse (consisting of about ten oscillations) on one side of the slab and the receiving transducer on the other side. For ultrasound, a detector measures both the amplitude and the phase of the wave, and it is thus relatively simple to phase-average the transmitted sound waves over a plane parallel to the slab sample. No significant distortion was observed in the shape of the coherent transmitted pulse. By matching the peaks of the transmitted and incident pulses, the group velocity was determined. The phase velocity was measured from the propagation time of the individual oscillations near the center of the pulse, for different sample thicknesses.

Results are shown in Fig. 4.7. Two features of the velocities can be noted. First, both the phase and the group velocity exhibit significant frequency dispersion. This is clearly a feature of resonant scattering, as neither of the component materials possesses these characteristics. Second, the longitudinal sound velocity of glass is 5.6 km/s, and the shear wave velocity is 3.4 km/s. Both are off the scales of Fig. 4.7. More importantly, the sound velocity in water is about 1.5 km/s, which means that in some frequency ranges the measured velocities are lower than the lowest velocity of the material components. This is generally impossible in the coherent potential approximation (CPA) for the effective material properties of composites, accurate in the limit of low frequency ($\omega \to 0$) [133]. The only plausible explanation is that the waves have been multiply scattered. However, such an explanation leads to a dilemma, since the measured velocities are supposed to be those of the coherent component.

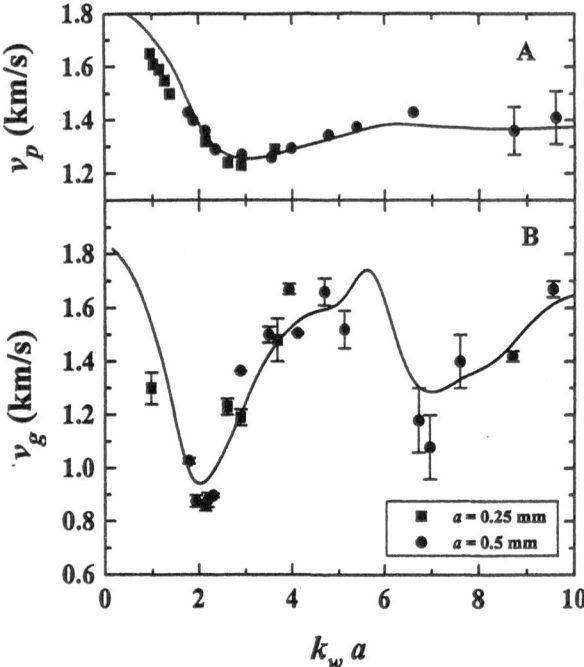

Fig. 4.7. Phase and group velocity of the coherent component of multiply scattered acoustic waves [reprinted with permission from Page, Sheng, Schriemer, Jones, Jing and Weitz, Science **271**, 635 (1996). © 1996 American Association for the Advancement of Science]

The theoretical component dealt with the problem of how to calculate the collective excitation modes of a system in the intermediate-frequency regime, defined as the spectral region where $\lambda \approx d$, where λ denotes the wavelength and d the size of the scatterers. The simplest way to explain the scheme is by starting with the scalar wave equation,

$$\left(\nabla^2 + \frac{\omega^2}{v_0^2}\right) \varphi(\mathbf{r}) + \omega^2 \left[v^{-2}(\mathbf{r}) - v_0^{-2}\right] \varphi(\mathbf{r}) = 0 \,, \tag{4.2}$$

where $\varphi(\mathbf{r})$ denotes the acoustic wave amplitude, and $v(\mathbf{r})$ the local phase velocity; $v_0(\mathbf{r})$ is a dummy variable, defined as a reference from which the inhomogeneous deviations are defined. In the frequency and wave vector representation, the configuration-averaged Green's function associated with (4.2) is given by

$$G(\omega, \mathbf{k}) = \frac{1}{\omega^2/v_0^2 - k^2 - \Sigma_{v_0}} \,, \tag{4.3}$$

where the self-energy Σ_{v_0} represents all the effects of multiple scattering caused by the inhomogeneities, given by the second term of (4.2). A physical way to define the dispersion relation is by looking at the peaks of the Green's function of the medium in the $(\omega, \mathbf{k})$ plane. Since Re Σ_{v_0} is always small at the peak, it can be combined with the first term of the denominator of (4.3) to re-define v_0. It makes sense to use the peaks of the spectral function, $-\text{Im}\, G(\omega, \mathbf{k})$, to define the dispersion relation in a strongly scattering medium. This is especially the case here since this definition coincides exactly with the CPA condition in the low-frequency limit [133].

The actual calculation of the spectral function has three notable features. First, the freedom in choosing the dummy variable was invoked by letting $v_0 = \omega/k$. Then,

$$-\text{Im}\, G(\omega, k) = \text{Im}\, \frac{1}{\Sigma_{\omega/k}} \,. \tag{4.4}$$

Second, instead of going to higher orders in scatterer density in the perturbational calculation of Σ, the actual calculation was done only to first order, but for a slightly more complex scattering unit which takes into account the geometric correlation between a glass sphere and the surrounding water, consisting of a glass sphere coated with a layer of water. The layer thickness is determined by the volume fraction of the spheres. Third, the elastic wave equation was used for the solid-sphere part, rather than the much simpler scalar wave equation.

The results are shown in Fig. 4.8, in which the peak of the spectral function is delineated by a white line. For comparison with the experimental results, the phase and group velocities were obtained in the usual way by treating the peak of the spectral function as the dispersion curve. These are shown as solid curves in Fig. 4.7. Excellent agreement is seen. There are no adjustable parameters in the theory.

The physical picture that emerges from this work is that when the scattering is strong, each scatterer must sense the scattered waves from the other particles, thus leading to the renormalization of the embedding medium. The amount of this renormalization depends on the strength of the scattering: the larger the scattering, the larger the adjustment. Moreover, it is physically plausible that as the concentration of scatterers increases, the material properties of the renormalized effective medium approach those of the scatterers themselves. Consequently, the individual scattering resonances inevitably become leaky and weakened, as the contrast between a scatterer and the embedding medium diminishes. It is precisely this effective renormalization of a strongly scattering medium that is sensed by the coherent group velocity. The procedure reported here identifies the frequencies and wave vectors of the minima in the coupled scattering resonances, and defines them to be the dispersion curve of the medium.

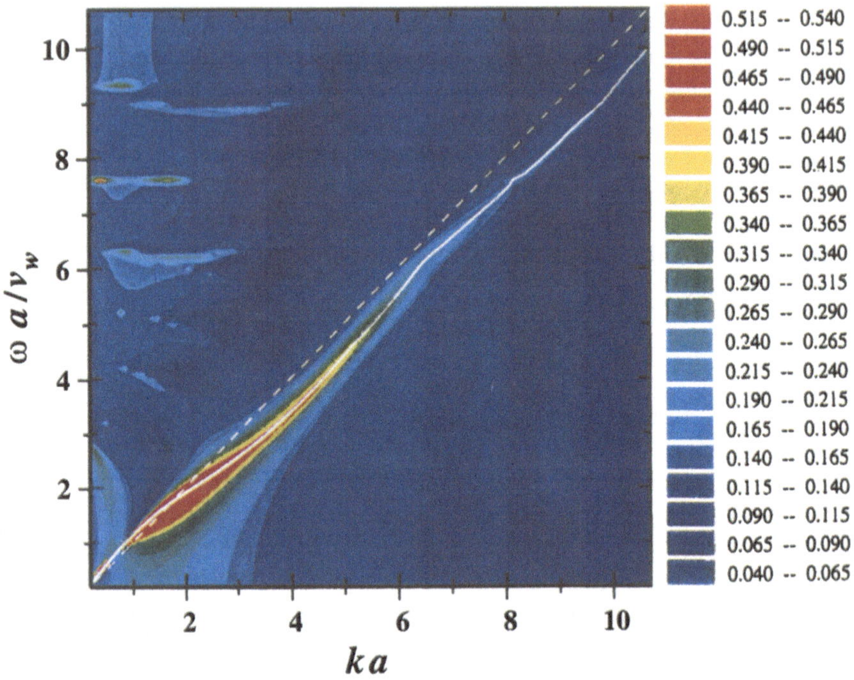
0.515 -- 0.540
0.490 -- 0.515
0.465 -- 0.490
0.440 -- 0.465
0.415 -- 0.440
0.390 -- 0.415
0.365 -- 0.390
0.340 -- 0.365
0.315 -- 0.340
0.290 -- 0.315
0.265 -- 0.290
0.240 -- 0.265
0.215 -- 0.240
0.190 -- 0.215
0.165 -- 0.190
0.140 -- 0.165
0.115 -- 0.140
0.090 -- 0.115
0.065 -- 0.090
0.040 -- 0.065
ω a/v_w
10
8
6
4
2
2
4
6
8
10
ka

Fig. 4.8. Spectral function $S(\omega, k)$ in the frequency–wavenumber plane, defined in (4.4), for acoustic waves in a random medium [reprinted with permission from Pagé, Sheng, Schriemer, Jones, Jing and Weitz, Science **271**, 635 (1996). © 1996 American Association for the Advancement of Science]

4.3 Transmission and Reflection of Acoustic Pulses by Randomly Layered Media

Jean-Pierre Fouque

An acoustic pulse entering a randomly layered slab and scattered by the random inhomogeneities will produce a reflected signal and a transmitted pulse, the asymptotic tail of which is referred to as a coda. In the regime of separation of scales (implying that the correlation length of the random inhomogeneities is much smaller than the typical wavelengths contained in the pulse, which are in turn much smaller than the scale of the medium, of the order of the size of the slab) an asymptotic analysis of the governing stochastic equations is possible and leads to a precise determination of the asymptotic probability distributions for the reflected and transmitted signals. The random fluctuations in the medium need not be weak. Moreover, it is possible to relate the local power-spectral densities of the reflected signal to the large-scale variations of the medium; this is done through a system of hyperbolic transport equations and leads to a solution of the inverse problem which consists of recovering these large-scale variations from the reflected signals.

As an illustration, we consider an acoustic wave traveling in a one-dimensional random medium located in the slab $0 \le x \le L$ and satisfying the linear conservation laws

$$\rho(x)\frac{\partial u}{\partial t}(x,t) + \frac{\partial p}{\partial x}(x,t) = 0,$$

$$\frac{1}{K(x)}\frac{\partial p}{\partial t}(x,t) + \frac{\partial u}{\partial x}(x,t) = 0, \tag{4.5}$$

where $u(x,t)$ and $p(x,t)$ are the velocity and pressure respectively of the wave, and $\rho(x)$ and $K(x)$ are the density and bulk modulus of the medium, admitting the following representation:

$$\rho(x) = \rho_0(x)\left[1 + \eta\left(\frac{x}{\varepsilon^2}\right)\right],$$

$$K^{-1}(x) = K_0^{-1}(x)\left[1 + \eta\left(\frac{x}{\varepsilon^2}\right)\right]. \tag{4.6}$$

Here ρ_0 and K_0, varying on the scale of the slab, represent deterministic parameters of the medium; $\eta(x/\epsilon^2)$ and $\nu(x/\epsilon^2)$ vary on the small scale ε^2 (the correlation length of the random medium) and represent the stationary-centered random fluctuations of the medium.

By defining the deterministic acoustic impedance as $I_0(x) = [\rho_0(x)K_0(x)]^{1/2}$ and the deterministic acoustic velocity as $c_0(x) = [K_0(x)/\rho_0(x)]^{1/2}$, we introduce the right-going wave $A(x,t)$ and the left-going wave $B(x,t)$, defined by

$$A = I_0^{-1/2}p + I_0^{1/2}u,$$

$$B = -I_0^{-1/2}p + I_0^{1/2}u. \tag{4.7}$$

These satisfy a hyperbolic system of first-order equations, with random coefficients. We write the boundary conditions corresponding to a pulse entering the slab from the left at time $t = 0$ and no wave entering the slab from the right:

$$A(0,t) = \varepsilon^{-\gamma}f(\frac{t}{\varepsilon}),$$

$$B(L,t) = 0, \tag{4.8}$$

where f is the pulse shape, $f(t/\varepsilon)$ containing wavelengths of order ε (larger than the correlation length ε^2 and smaller than the size of the slab); $\varepsilon^{-\gamma}$ is the amplitude of the incoming pulse, the energy entering the slab being of order $\varepsilon^{1-2\gamma}$; $\gamma = 1/2$ corresponds to a fixed energy and $\gamma = 0$ corresponds to fixed amplitudes. The transmitted and reflected waves respectively are given by $A(L,t)$ and $B(0,t)$.

4.3.1 Transmitted Pulse

The analysis of the transmitted pulse (coherent transmitted front) has been carried out in [230] and [231] for the one-dimensional case. Introducing the deterministic travel time, defined by

$$\tau_0(x) = \int_0^x \frac{dy}{c_0(y)}, \tag{4.9}$$

the transmitted pulse, viewed on the scale ε of the initial pulse, is given by

$$A[L, \tau_0(L) + \varepsilon\sigma], \tag{4.10}$$

where σ is the time in the ε-scale.

In the regime of constant amplitude ($\gamma = 0$), the probability distribution of this transmitted pulse converges, as ε goes to 0, to the probability distribution of the initial pulse, spread by a convolution with a Gaussian density and shifted by a Gaussian random variable. The variances of both the Gaussian density and the Gaussian shift depend upon the size of the slab (distance of propagation) and the covariance of the random medium.

This analysis has been generalized [232] to a 3D randomly layered medium: in the same regime of separation of scales, the asymptotic probability distribution of the pressure field at the bottom of the slab generated by a pulse at the surface is obtained; the main difference from the one-dimensional case is a convolution with the derivative of a Gaussian density instead of a Gaussian density itself, only one Gaussian variable being needed to describe the distribution of the field at various offsets.

These results are an extension of the O'Doherty–Anstey theory [233], which asserts that for weak fluctuations, which in fact we did not assume, the traveling pulse retains its shape except for a small spreading (convolution by a Gaussian density), and is deterministic when observed by an observer traveling at the same random velocity as the wave ($c(x) = [K(x)/\rho(x)]^{1/2}$), while it is stochastic (shifted by a Gaussian random variable) when the observer's velocity is the deterministic velocity of the wave $c_0(x)$.

4.3.2 Reflected Signal

The reflected signal $B(0, t)$ obtained from a pulse scattered by a randomly layered medium in the regime of separation of scales described above has been studied by Papanicolaou et al. The one-dimensional case has been treated in a series of papers [234–237], the last paper being devoted to the one-dimensional inverse problem. The 3D layered case is introduced in the review paper [238], as well as the corresponding inverse problem. A generalization to elastic waves can be found in a recent paper [239].

The first observation of this study is that, apparently, the information about the large-scale variations of the medium is lost in the noise present in the reflected signal. This noise is due to the multiple scattering of the pulse as it propagates into the medium. One of the main ideas of the theory is that the local properties of this nonstationary random signal contain information. More precisely, locally around a time t_0 and on the scale of the initial pulse ε (large compared to the correlation length ε^2 of the inhomogeneities but small compared to the distance of propagation), this signal, given by $B(0, t_0 + \varepsilon\sigma)$, is asymptotically (as ε goes to 0) stationary, centered and normally distributed. The spectral density of the asymptotic stationary Gaussian process is related to the deterministic part of the coefficients of the medium ($\rho_0(x)$ and $K_0(x)$) through a hyperbolic system of transport equations (the so-called W-equations). This is obtained by an invariant embedding method and an asymptotic analysis (in the white-noise approximation regime) of the differential equations with random coefficients describing the wavefield. Tools such as convergence of stochastic processes or stochastic calculus with respect to Brownian motions are needed in this study. In the case of a uniform background (no large-scale variations of the medium), this spectral density can be computed explicitly; this is very important for comparison to various simulations [238].

The inverse problem, which consists of the reconstruction of the large-scale variations of the medium appearing as coefficients in the underlying transport equations, requires good statistical estimators for the local covariances of the reflected signals, namely the mathematical expectation of $B(0, t_0)B(0, t_0 + \varepsilon\sigma)$. This has been done by using a windowed Fourier transform [238]. An estimate based on a wavelet transform [240] has been proposed, as well as an estimate for the small parameter ε representing the separation of scales, i.e. the ratio of the size of the inhomogeneities to the wavelengths or the ratio of the wavelengths to the distance of propagation.

In the context of ultrasound it is possible to use time-reversal mirrors to construct very robust and convergent estimators [241]. A time-reversal mirror is a device developed at the ESPCI-Paris by Fink et al. [242]; it is described in Sect. 4.1.3. Using this method, it has been shown [241] that one obtains a new reflected signal which becomes asymptotically deterministic and whose Fourier transform is nothing more than the spectral density. The medium itself is doing the computation of the local covariance in the time domain and therefore produces the best estimator one can wish for.

The theory for the reflected signal briefly presented here is one-dimensional (or 3D layered) and it is not clear at all what happens in its multidimensional generalization.

5. Lidar

François Nicolas and Pierre Flamant

Lidar is an acronym for "light detection and ranging", a term used to describe profiling systems for the atmosphere based on laser devices. Lidars operate in the optical frequency domain in a fashion similar to radars in the microwave domain. The atmospheric parameters retrieved by lidars deal with the microphysics of clouds (densities, sizes and distributions), the chemical composition of the atmosphere (densities) and the dynamics (wind velocity and turbulence strength).

The various lidar techniques are named after the light–scatterer interaction processes which take place, i.e. (elastic) backscatter lidar, Raman lidar, differential absorption lidar (or DIAL), Doppler lidar, etc.; the parameter under study (e.g. cloud, aerosol, water vapor, ozone, wind, etc.); and/or the detection process.

In lidar the driver is its application to meteorology, climate and pollution. Its performances in terms of range and accuracy rely on atmospheric propagation, light–matter interaction processes and instrumental parameters. Obviously the first two call on developments in basic physics and mathematics. Sects. 5.3 and 5.4 review the present status of multiple-scattering contribution to cloud lidar returns and of coherent backscattering respectively. Beforehand, an introduction (Sect. 5.1) provides the background, while the principle of lidar is presented in Sect. 5.2.

5.1 Historical Background

Soon after the discovery of the laser effect in ruby in 1960, the potential applications to external geophysics were recognized. The first significant results on remote sensing of cloud and clear air were reported in 1962–63. Since then, the field has expanded rapidly. Over the past thirty years, lidar technology has evolved considerably thanks to progress made in computer science, laser physics, optics and electronics. Today, applications call for sophisticated systems operating from the surface, from air platforms and, more recently, even from space. Over the years, emphasis has been put on the operationality, reliability and compactness of lidars for various applications.

Besides the many efforts directed to technical improvements, significant advances have been made dealing with direct and inverse problems such as laser propagation in a turbulent atmosphere and signal processing. The close connection between lidar physics, geophysical applications and the underlying basic physics clearly emphasizes the interdisciplinary aspect of the subject and provides clear links to the topics presented elsewhere in this book.

The current objective is to develop advanced theories leading to the development of new lidar methodologies and/or new tools to achieve better or new types of measurements. To illustrate this, we consider the laser and the propagation of scattered light in the atmosphere. Most of the work in the past thirty years was based on the single-scattering approximation, even though it does not hold in dense media such as clouds. Multiple scattering then plays a major role in the extinction and depolarization of the scattered light. So far, multiple scattering is usually treated as a first-order perturbation and only a small amount of work has dealt with a complete theory. At this stage, it is worth remembering that the direct problem is only important as a tool to solve the inverse problem of retrieving atmospheric parameters.

A difficulty in lidar as it stands, for any optical system, is its a priori inability to penetrate dense clouds with large optical depths. However, partial cloud cover and/or optical porosity on a small scale allow one to study multilayer clouds. Also, the occurrence of multiple scattering allows one to penetrate deeper in dense media than is possible according to single scattering.

5.2 Principle

The lidar principle is similar to the radar principle: a collimated laser beam is sent into the atmosphere and it interacts with molecules or particles; a fraction of the laser energy is scattered at π radians (or backscattered) and collected by a telescope (Fig. 5.1). A pulsed emission is required for a ranging capability beyond a few hundred meters (up to 100 km) and for range-resolved measurements. The lidar range $R = ct/2$ is linked to the round-trip time of flight t and the speed of light c as usual in the single-scattering approximation (see later). The range resolution $\Delta R = c\tau/2$ is limited by the pulse duration τ (and/or the response time of the detection system). Because lidar measurements are taken along a line of sight, several measurements are required to sample the atmosphere in two or three dimensions, by scanning the lidar line of sight.

The light–scatterer interaction provides both the signature of the parameter to be retrieved and the strength of the lidar signal to be processed. The basic interactions of particles and molecules with light are elastic scattering, vibrational and rotational spontaneous Raman scattering, absorption and fluorescence. It is worth noticing that the Doppler effect is always involved in these processes. It may be used for velocity measurements. Lidars have been used with a great variety of laser sources, covering the whole optical

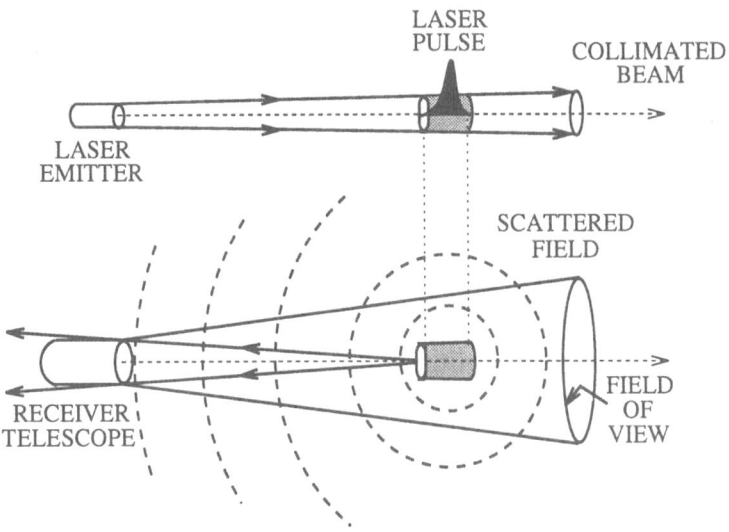

Fig. 5.1. The principle of lidar remote sounding in the atmosphere: a laser pulse is emitted into the atmosphere. A fraction of the scattered energy (*dotted lines*) comes back and is collected by a telescope

spectrum from the near ultraviolet to the mid-infrared, but nowadays only reliable laser sources (i.e. Nd-YAG, XeCl, CO_2, etc.) find applications.

The lidar signal $S(R)$ is the scattered optical power collected by the telescope, or the output current after detection and analog–digital conversion. Assuming single scattering, it is given by

$$S(R) = K \frac{\beta(R)}{R^2} \exp\left(-2 \int_0^R \alpha(z)\, dz\right). \tag{5.1}$$

K is an instrumental constant which includes the laser energy per pulse, the receiver's collecting area (A), the transmitter–receiver's optical efficiency, the detector quantum yield and an appropriate electronic amplification factor. In (5.1), the atmosphere is completely described by two coefficients, namely the backscattering coefficient $\beta(R)$ and the extinction coefficient $\alpha(R)$. Assuming a probing laser wavelength of 0.53 μm, an energy of 50 mJ and a collecting area of 0.1 m^2, the optical power backscattered by clear air at 3 km is 10^{-8} W.

When the same scatterers are responsible for extinction and backscattering, $\alpha(R)$ and $\beta(R)$ are related by $\beta(R) = \omega(R)\, p(\pi, R)\, \alpha(R)$, where $\omega(R)$ is the single-scattering albedo and $p(R)$ is the scattering phase function at range R. The inversion of (5.1) usually assumes that this relation holds with a proportionality factor $\omega(R)\, p(\pi, R)$ independent of the range.

The coefficients α and β vary by several orders of magnitude depending on the probing laser wavelength and the meteorological conditions (i.e. the presence of clear air or clouds). In clear air, the λ^{-4} dependence for Rayleigh scattering generates a six-order-of-magnitude variation for wavelengths ranging from 0.3 μm to 10 μm. Similarly a λ^a dependence for Mie scattering, where $a \approx -(2 \pm 1)$, applicable to scatterers with sizes of the order of or larger than the probing laser wavelength, implies several orders of magnitude of variation. In practice β ranges from 10^{-10} to 10^{-3} m^{-1} sr^{-1}, and α from 10^{-5} to 3×10^{-2} m^{-1}, corresponding to scattering mean free paths ℓ between 0.03 km and 100 km. These ranges give rise to quite distinct propagation regimes (i.e. single scattering, double and multiple scattering, and the diffusion approximation). Such huge variations are very important in lidar design and performance.

Direct detection is used for power measurements ($|E|^2$, where E is the scattered electric field), while heterodyne detection is used for amplitude measurements (E) in order to determine the Doppler frequency shift. Heterodyne detection calls for a high degree of partial transverse coherence of the scattered light at the receiver in order to achieve an efficient optical mixing with the emission of a reference CW laser (the local oscillator). The (partially) coherent illumination gives rise to speckle effects, which are responsible for fluctuations in the detected signal. These fluctuations then follow a Rayleigh (exponential) law. On the other hand, when direct detection is used, the fluctuations follow a narrow Gaussian law because the number of speckles at the detector is of the order of several hundred and the detector averages the fluctuations.

The propagation of the lidar pulse and the backscattered light through the atmosphere may occur in clear air characterized by a refractive index turbulence C_n^2, or in clouds. Clouds are colloidal media made of air molecules and particles; the two types of scatterers can be mixed or distributed in layers at different altitudes. The scattering volume giving rise to the lidar signal is determined by the laser pulse; it is a few cubic meters or more. It follows that the total number of scatterers (molecules and/or particles) is tremendous. As a result, the density fluctuations of the medium are negligible and the scattering volume is a disordered medium rather than a random medium.

5.3 Incoherent Multiple Scattering

Lidar methodologies were first developed relying on the single-scattering approximation. However, when experiments in clouds were conducted, it was realized that multiple scattering contributed significantly to lidar signals. Indeed, Pal and Carswell [243] measured a cross-polarized signal in lidar returns from water clouds, in a case where the scatterers were known to be spherical and thus not to depolarize for scattering at π radians. The second piece of evidence of multiple-scattering contributions in cloud lidar returns was obtained by Allen and Platt [244]: they could directly measure the multiple-scattering signal by removing the single-scattering component with a center-blocked field stop, which restricted the receiver field of view to a region outside the transmitted beam.

5.3.1 Single-Backscattering Approximation

In the case of lidar cloud returns we can give a simple account of multiple scattering. The scatterers in clouds have dimensions greater than the laser wavelength, e.g. of the order of microns in water clouds, hundreds of microns in ice-crystal clouds and millimeters in precipitating clouds. Hence the scattering phase functions are highly peaked in the forward direction mainly because of Fraunhofer diffraction. The width of the diffraction peak is characterized by the so-called diffraction angle $\theta_d = \lambda/2r$, where r is the characteristic radius of the scatterers.

As long as the two-way propagation length is smaller than the transport mean free path ℓ^*, or equivalently the optical depth τ is smaller than $0.5/(1-g)$, where g is the asymmetry factor, the propagation remains anisotropic in the cloud. Therefore, multiple scattering can be described within what has been called the single-backscattering approximation [245]. As shown in Fig. 5.2, a light path can comprise several scatterings in the forward direction, on the way either to or from the cloud, but a single scattering near π.

Let $\theta_c = (L/R - L)\,\theta_{FOV}$, where R is the measurement range, L is the altitude of the cloud base, and θ_{FOV} is the angular field of view of the lidar. Photons scattered at angles smaller than θ_c remain within the receiver field of view up to the range R and hence affect the attenuation of the signal. As the cloud properties are mainly determined by attenuation, we will consider multiple scattering to be present as soon as there is a modification in the

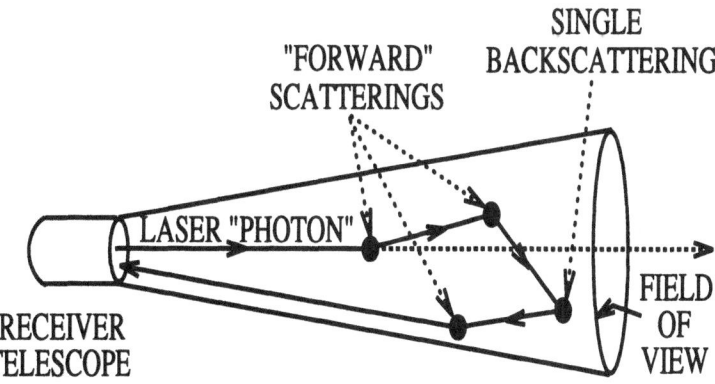

Fig. 5.2. Illustration of the single-backscattering approximation: a light ray contains several forward scatterings on its way to and from the cloud, but only one scattering near π, i.e. one backscattering

Table 5.1. Lidar propagation regimes in clouds.

$\tau \lesssim 0.5/(1-g)$		$\tau \gg 0.5/(1-g)$
Single backscattering		Diffusive regime
$\xi \ll 1$	$\xi \sim 1$	
Single scattering	Multiple scattering	

attenuation (or the derivative) of the signal, and not in the signal itself, as would have been the case with the usual definition of multiple scattering.

With this new definition, the nondimensional parameter measuring the importance of multiple scattering is the ratio ξ of the two characteristic angles defined above:

$$\xi = \frac{\theta_c}{\theta_d}. \tag{5.2}$$

Table 5.1 lists the various lidar regimes with respect to the optical depth τ and the parameter ξ.

The single-backscattering approximation allows one to write the lidar signal in the following form [245]:

$$S(R) = K \iiint F(R, \mathbf{r}, \mathbf{n}) \left[\omega(R)\alpha(R) p(R, \mathbf{n} \rightarrow \mathbf{n}') \right]$$
$$\times B(R, \mathbf{r}, \mathbf{n}') \, d\mathbf{r} \, d\mathbf{n} \, d\mathbf{n}', \tag{5.3}$$

where three distinct parts of the photon path clearly appear:

1. $F(R, \mathbf{r}, \mathbf{n})$ is the radiant intensity at $(R, \mathbf{r})$ in the direction $\mathbf{n}$, due to propagation from the laser source;
2. $\left[\omega(R)\alpha(R) p(R, \mathbf{n} \rightarrow \mathbf{n}') \right]$ is the probability of scattering at an angle $(\mathbf{n}, \mathbf{n}',$ i.e. the backscattering probability;
3. $B(R, \mathbf{r}, \mathbf{n}')$ is the probability for a photon located at $(R, \mathbf{r})$ directed along the direction $\mathbf{n}'$ to be collected by the receiver.

The following three properties of the single-backscattering regime imply that the use of lidars to monitor cloud properties remains very attractive in spite of multiple scattering. First, the ranging ability of lidar is maintained because of a negligible temporal spreading of the pulse. This has been demonstrated experimentally with laboratory-scale measurements [246] and with real clouds [247]. Second, because Fraunhofer diffraction is independent of polarization, the depolarization observed in lidar returns arises mainly at the backscattering event. The depolarization signal can thus be related to the properties of the single-scattering phase function near π. Third, it can be shown that the average Doppler shift is not affected by multiple scattering, and that the spectral broadening of the backscattered field remains dominated by the atmospheric wind fluctuations occurring at the backscattering event [248]. Thus, wind velocity measurements in clouds can be performed with Doppler lidars and the performance is not degraded by the presence of several forward scatterings.

It is important to keep in mind the limitations of the single-backscattering approximation. In effect, if the penetration depth is greater than the transport mean free path ℓ^*, the propagation becomes diffusive and photons can suffer several wide-angle scatterings, leading to long tails in the returns, like the coda in seismology. To detect such a diffusive signal, the lidar field of view must be such that its footprint is of the order of the transport mean free path ℓ^*. This condition has been met during the shuttle-based "Lidar In Space Technology Experiment" (LITE) [249], where the footprint was as large as several hundred meters, and abnormally long delay times characteristic of diffusive scattering have been reported when low-altitude marine cloud prevailed [250]. Recently, Davis et al. proposed to use this diffusive regime by increasing the receiver field of view, i.e. by working with so-called wide-angle imaging lidars (WAILs) [251]. The retrieved parameters would in this case be an average geometrical thickness and an average transport mean free path.

5.3.2 Direct Problem

The direct problem of calculating the lidar returns for a given set of cloud and instrumental parameters has been an extensive field of research since the first experimental evidence for multiple scattering was obtained in the seventies. A priori, it would suffice to solve the radiative transfer equation [1, 2, 76] with appropriate initial and boundary conditions relevant to the lidar system. However, this problem has not been solved exactly, because of the difficulty of taking into account the backscattering geometry. Recently, a concerted modeling effort was initiated by the international cooperative group MUSCLE (multiple scattering lidar experiments), which publishes annual reports. In a comparison study published in a feature issue of *Applied Physics* B [252], the MUSCLE participants applied their models and algorithms to a common problem (Fig. 5.3).

This study provides a good picture of the state of the art for the various methodologies for solving the direct problem. Two types of approaches were presented, namely numerical simulations and analytical models.

Numerical simulations are based on a Monte Carlo procedure. In such simulations, the light beams are discretized into a set of "photon" trajectories along which photons propagate in the cloud according to the probability laws of scattering. Ideally, the signal would thus be built up by adding the trajectories ending in the detector. However, these trajectories are very unlikely to occur and some tools, or variance-reduction methods, have to be used to make the calculation time reasonable. The most natural method is called the semianalytic procedure [253]: it consists of calculating analytically at each scattering step the probability that a photon directly reaches the detector. To increase computation efficiency, more sophisticated variance-reduction methods have also been proposed, e.g. by Starkov et al. [254], who used nonphysical probability distributions constructed from a priori information. The advantage of Monte Carlo simulations is that they can deal with actual in-

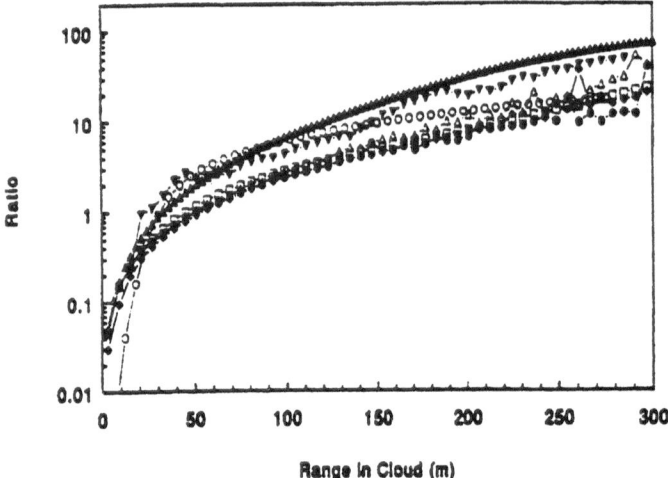

Fig. 5.3. Calculated ratios of the multiple to single scattering contributions to the lidar return from a uniform C.1 cloud for a receiver field of view of 10 mrad; the cloud base is at 1 km and the lidar is directed vertically (▲, DREV's model; ●, Florence group's model; ▼, Israel group's model; ♦, Munich group's model; △, NASA group's model; ○, Swiss group's model; □, Minsk group's model) [reprinted from Bissonnette, Bruscaglioni, Ismaelli, Zaccanti, Cohen, Benayahu, Harach, Cohen, Flesia, Schwendimann, Oppel, Winkel, Zege, Katsev and Polonsky, Appl. Phys. B **60**, 355 (1995). © 1995 Springer-Verlag]

strumental parameters, cloud inhomogeneity and stratification. They now compare very well with each other and have been validated experimentally [255, 256]. Thus, they can really serve as numerical reference experiments.

A variety of analytical expressions have been proposed to calculate lidar returns; except for one, they all assume the single-backscattering approximation to be valid. Bruscaglioni and Ismaelli [257] and Eloranta [258] first derived analytical expressions for double scattering. The latter work was then extended to multiple scattering by Eloranta and Shipley [259] under the assumption of a Gaussian shape for the forward peak of the phase function. A more phenomenological approach was suggested by Bissonnette [260], who represented the lateral beam spreading by a diffusion process, described by the radiative transfer equation. To increase the accuracy of this model, Bissonnette separated the multiple-scattering processes, i.e. no scattering, single scattering and multiple scattering on the ways to and from the cloud, and gave a phenomenological expression for each [261]. Flesia and Schwendimann [262] extended the Mie theory to multiple scattering by calculating the n-fold-scattered electromagnetic field.

The most interesting work is that one proposed by Zege et al. [263] and Katsev et al. [264]: the optical reciprocity theorem and the aspect invariance property [265] lead them to simplify (5.3) to

$$S(R) = K \int \omega(R)\alpha(R)p^{b}(R, |\mathbf{n}_{\perp}|)\, H(R, \mathbf{n}_{\perp})\, d\mathbf{n}_{\perp}, \tag{5.4}$$

where p^{b} is the backscattering phase function and $\mathbf{n}_{\perp}$ and $\mathbf{n}'_{\perp}$ are projections onto the plane perpendicular to the lidar axis.

These authors show that the function H of (5.4) is given by

$$H(R, \mathbf{n}_{\perp}) = \varphi_{\text{eff}}(R, \mathbf{r} = 0, \mathbf{n}_{\perp}), \tag{5.5}$$

where $\varphi_{\text{eff}}(R, \mathbf{r} = 0, \mathbf{n}_{\perp})$ is the radiant intensity due to transmission from an effective source φ_{eff} in an effective medium made of the same scatterers but with twice the density. The effective source is defined by the following convolution of the laser source function φ_{las} and the so-called "back-propagated receiver source" φ_{rec} [270]:

$$\varphi_{\text{eff}}(0, \mathbf{n}_{\perp}) = \int d\mathbf{r} \int d\mathbf{n}'_{\perp} \varphi_{\text{las}}(0, \mathbf{r}, \mathbf{n}'_{\perp})\, \varphi_{\text{rec}}(0, \mathbf{r}, \mathbf{n}'_{\perp} + \mathbf{u}_{\perp}). \tag{5.6}$$

This approach thus assumes the single-backscattering approximation to be valid and the phase function to be independent of the azimuthal scattering angle, but there is no assumption about the shapes of the laser and receiver profiles or about the shape of the phase function; in particular, the diffraction and the ray-tracing contributions to the forward peak are both taken into account, and nothing is assumed about the shape of the phase function near backscattering.

5.3.3 Inverse Problem

Many techniques have thus been developed to solve the direct problem. However, researchers are interested in the parameters they can retrieve from an experimental lidar signal, that is to say the inverse problem. Traditionally, the simple description of the lidar signal proposed by Platt [266] is used: it consists of replacing the scattering coefficient in the single-scattering lidar equation by some effective coefficient accounting for multiple scattering. The use of effective coefficients has also been discussed by Nicolas et al. [245], and it has been shown to be exact in some cases of interest.

The simplicity of this approach makes it very attractive in the context of inversion. In this case, multiple scattering is often seen as a drawback. However, the development of the above-mentioned analytical models and simulation codes makes it now possible to use multiple scattering as a tool to gather new information on microphysical properties (i.e. particle effective radius and liquid water content). This can be achieved by the use of a multiple-field-of-view (MFOV) lidar as proposed by Bissonnette and Hutt [267]. Such lidars image the transverse spreading of the beam, which is due to multiple scattering; the width of this spreading is related to the width of the diffraction peak, and hence to the size of the scatterers.

Figure 5.4 shows an example of lidar signals obtained at the Defense Research Establishment in Valcartier (DREV) with an MFOV lidar [268].

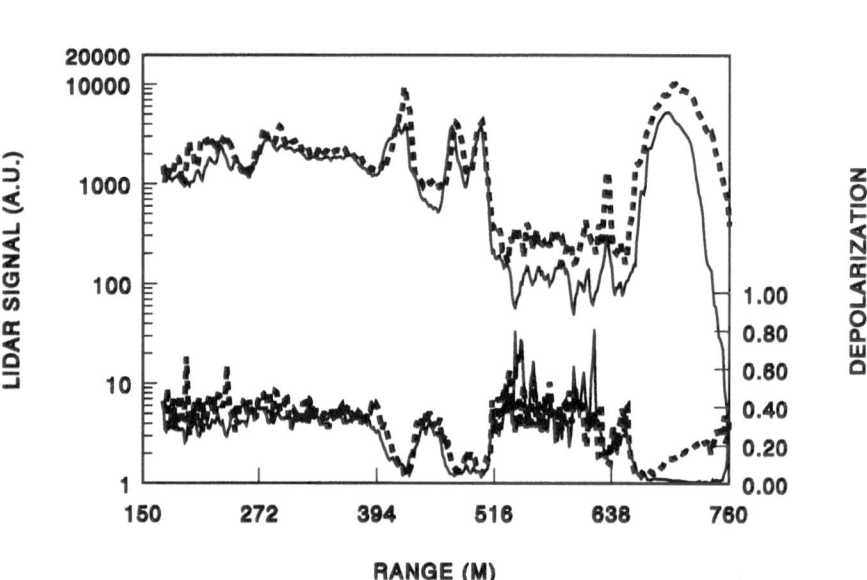

Fig. 5.4. Lidar returns at two field of views obtained with the MFOV lidar at Defense Research Establishment in Valcartier (DREV). Several layers of water clouds can be distinguished among the snow [reprinted from Roy and Bissonnette in: Advances in Atmospheric Remote Sensing with Lidar: Proc. 18th Radar Conference 1996, edited by A. Ansmann et al., with kind permission from the authors. © 1996 Springer-Verlag]

The upper two curves are the intensity returns from fields of view with half-angles of 0.6 and 6 mrad, while the lower two curves are the depolarization ratios for these two fields of view. In this figure, several atmospheric layers can be distinguished: the presence of some snow can be identified from 150 to 390 m because the lidar signal is almost constant, with a high depolarization ratio ($\approx$ 40 %) characteristic of hexagonal crystals. Two thin water-droplet layers then appear at 400 m and at 480 m, and they are associated with low depolarization ratios. Finally, the main water-droplet cloud is present above 650 m: the depolarization ratios drop almost to 0, but since the layer is thick we note that the 6 mrad depolarization ratio rises again, now because of the presence of multiple scattering. Thus the mere qualitative analysis of MFOV and polarization lidar signals provides us with some information on the structure of the atmosphere. The next step is to use the difference between the returns in the two (or more) fields of view to retrieve the size of the scatterers. Preliminary inversion results can be found in [261] for water clouds and in [269, 270] for ice clouds; nevertheless, the achievement of a stable and robust inversion scheme remains one of the open issues in this field.

5.4 Coherent Backscattering

The relevance of the enhancement due to coherent backscattering in lidars has been studied by Nicolas [270] and Borovoi [271] with a phenomenological analysis. Their conclusions were that for lidars using direct detection, no enhancement shows up. On the other hand, for heterodyne lidars the enhancement factor is almost maximum, i.e. close to two, the value known in optics. In the latter case, the laser profile is identical to the telescope profile and they are both limited by diffraction. Thus, to any multiply scattered photon there corresponds a time-reciprocal one, leading to the factor of two.

6. Applied Mathematics

Jean Lacroix

The study of random Hamiltonians has become an important topic in applied mathematics. General questions that can be posed are "what is localization?" [272], particularly in relation to the spectrum of the Hamiltonian and to the dynamics of a wave packet, and "under what conditions does localization occur?", that is, for what energies, for what amount of disorder and for what systems? A lot of progress has already been made. The main result, proved under certain conditions, is the "pure point nature" of the spectrum of time-independent Schrödinger Hamiltonians, which yields localization for the solutions of the Schrödinger equation. The proof that states are localized for "almost all" realizations of the disorder in one-dimensional systems was first given some thirty years ago [273], following a physical argument by Mott and Twose [274], and modified and extended later, with important contributions by Kunz and Souillard [275] and Delyon et al. [276]. The result is a, by now, famous Thouless formula relating the decay of the wave function (the Lyapunov exponent) to the density of states in the random medium [277, 278]. Fröhlich and Spencer [279] formulated, using a new multiscale analysis, the first rigorous proof that localization occurs in the discrete Anderson model for "sufficiently large" disorder and at "sufficiently low" energy (that is, "sufficiently" far away from the band). This proof has recently been improved considerably by Aizenman and Molchanov [280].

Recent work addresses localization of acoustic and electromagnetic waves, in particular in periodic media with spectral gaps. Using a variant of the multiscale analysis developed by Fröhlich and Spencer, the existence of localized eigenstates in the vicinity of the edges of the gap can be established. A key ingredient is a new Wegner-type estimate for a class of operators with off-diagonal disorder [281, 282]. A big challenge in the domain of applied mathematics is still the proof of localization in continuous models [283], and a rigorous proof of diffuse behavior (i.e. $\langle x^2(t) \rangle /t \rightarrow$ constant as $t \rightarrow \infty$) for small disorder.

6.1 Localization Theory

Spectral properties of second-order operators with random coefficients have been intensively studied over the last twenty years. Various definitions of localization have been considered in physical and mathematical papers and we intend to clarify their relationships. The Schrödinger equation

$$i\frac{\partial\psi}{\partial t} = H\psi\,,\tag{6.1}$$

where the operator H is given by $H = H_0 + \lambda V$, is certainly the most studied model. In general, the free Hamiltonian H_0 is the conventional Laplace operator, the multiplication operator V is viewed as an exterior potential and λ is some coupling constant. Under suitable assumptions H is a self-adjoint operator on the Hilbert space $\mathcal{H}$ of square-integrable functions over $\mathbb{R}^d$ (continuous case) or $\mathbb{Z}^d$ (lattice case). An element $\psi_0(x) \in \mathcal{H}$ is said to be a *bound state* if the function $\psi_t(x) = \exp(\mathrm{i}tH)\,\psi_0(x)$ exhibits the localized behavior

$$\lim_{r\to\infty}\sup_t \int_{|x|>r} |\psi_t(x)|^2\,\mathrm{d}x = 0\,,\tag{6.2}$$

and to be a *unbound state* if it has the "diffusive" behavior

$$\lim_{T\to\infty}\frac{1}{T}\int_{-T}^{T}\left(\int_{|x|\leq r} |\psi_t(x)|^2\,\mathrm{d}x\right)\,\mathrm{d}t = 0\,.\tag{6.3}$$

When the spectrum of H is pure point then any element of $\mathcal{H}$ is a bound state and H is said to be strongly localized. On the other hand, in the study of crystals with a periodic potential and an absolutely continuous spectrum with a band structure, any element of $\mathcal{H}$ is an unbound state. If this periodic potential is randomly perturbed then it turns out that under suitable assumptions for the strength of the randomness, the situation dramatically changes since, for almost all realizations of the random potential V, the spectrum of H becomes pure point. In order to give an idea of the mathematical approach to such a phenomenon we will consider essentially the simplest case, of Schrödinger operators on the one-dimensional lattice. The abbreviation "*a.s.*" will be used to mean "almost surely" or "for almost all realizations of the potential".

6.1.1 One-Dimensional Lattice

Assume that the random potential V is given by an independent sequence of nonconstant random variables $V(n)$ at each site $n \in \mathbb{Z}$, each with the same probability distribution (the so-called Anderson model). It is well known that the spectral properties of H are closely related to the asymptotic behavior of the solutions of the eigenvalue equation $H\psi = E\psi$, namely

$$(H\psi)(n) = -\psi(n-1) - \psi(n+1) + \lambda V(n)\psi(n) = E\psi(n).\tag{6.4}$$

It would be tempting to use perturbational arguments around the free case $\lambda = 0$ to obtain some spectral information, at least for small values of λ. Unfortunately such a procedure cannot be successful in general, since the spectrum is absolutely continuous in the free case but has been proved to be *a.s.* pure point for any $\lambda \neq 0$, as first suggested by the work of Anderson [3] and Borland [273]. This strong localization property is somewhat unexpected, as shown in the next section.

6.1.2 Strong Localization is Very Unlikely

Let the energy E and $\lambda \neq 0$ be fixed. Equation (6.4) is a second-order recurrence equation, and hence the solutions can be computed along the positive and negative directions of the Z-axis by means of two products of (random) 2×2 transfer matrices applied to some initial two-dimensional vector. The Lyapunov exponents of these two products can be shown to be equal and strictly positive. It follows (using Osseledec's theorem) that *a.s.* any solution of (6.4) grows exponentially in at least one direction of the Z-axis, and therefore *a.s.* the energy E cannot be an eigenvalue of H. This is a bad indication, since strong localization implies that the spectrum of H possesses a dense subset of eigenvalues. Nevertheless, a weaker result obtained by Ishii and Matsuda [284] and Pastur [285] is an immediate consequence of the positivity of the Lyapunov exponent: *a.s.* the operator H has no absolutely continuous spectrum. This does not automatically imply strong localization since it remains necessary to exclude the singular continuous spectrum; ergodic models exist without eigenvalues, for which the Lyapunov exponent is positive. This weaker form of localization often appears in wave propagation theory since it is relatively easy to prove that the exponential decay of the transmission coefficient in a disordered slab is just a consequence of the positivity of the Lyapunov exponent of the product of the random transfer matrices.

6.1.3 Strong Localization *a.s.* Occurs

Fortunately, it turns out that the set of potentials of full probability measure for which the energy E is not an eigenvalue of H has a very chaotic behavior with respect to the value of E, and it is impossible to construct such a set of positive probability measure corresponding to a whole interval of energies. The essential step in the proof of strong localization is to "translate" almost-sure properties obtained at a fixed energy into almost-sure properties of spectral measures of a given set of energies. Looking at the eigenvalue equation (6.4), which can be rewritten as $H_0\psi = (E - \lambda V)\psi$, it is in some sense equivalent to shift either the energy E or the potential V. This can be done, assuming that the probability distribution of V has good properties with respect to translations, in particular if this distribution is absolutely continuous. In this situation the so-called "Kotani's trick" [286], the perturbation method of self-adjoint operators of Simon and Wolff [287] and the Markov chain operators method introduced by Goldsheid, Molcanov and Pastur [288] and later used by Kunz and Souillard [275] immediately imply the strong localization property. Using some algebraic and operator machinery the same result can be obtained for quasi-one-dimensional systems or strips [289]. Unfortunately, attempts to solve the multidimensional case by means of strips whose width is allowed to grow to infinity have not yet been successful. Strong localization also holds for general distributions of potentials but the proof is much more involved, and this certainly indicates that a "good" proof of strong localization is still missing.

6.1.4 The Multidimensional Lattice Case

In higher dimensions one needs to replace the positivity of the Lyapunov exponent by the almost-sure exponential decay of the Green function at a given energy. This result has been obtained by Fröhlich and Spencer [279] for large disorder, that is, for large values of λ, or for large energies using the so-called "multiscale analysis". As in the one-dimensional case, this property implies the absence of an absolutely continuous spectrum. In the same regimes and assuming an absolutely continuous distribution of the potentials one gets a.s. strong localization by means of one of the two methods mentioned above. With the same assumptions, a simpler proof valid for a much wider class of Hamiltonians has been given by Aizenman and Molcanov [280]. A proof for a general distribution of potentials has been obtained by Fröhlich et al. [290] and von Dreifus and Klein [291].

6.1.5 Extensions

Up to now, strong localization in the one-dimensional case has been proved for independent potentials (Anderson model) and for potentials governed by a Markov process. In the continuous case, the first proof was given in 1977 by the "Russian school" [288] and was considerably improved and extended to many models by Carmona [292]. In higher dimensions this property has only been proved for large energies, large disorder or near the band edges of the spectrum of H_0 [293, 294], or for continuous models with Poisson disorder [283]. Nevertheless it is believed that localization always occurs in the two-dimensional Anderson model and that for higher dimensions, absolutely continuous spectra should exist at low energies. Other models, including acoustic and electromagnetic waves, have been considered and it has then been proved that strong localization occurs near the edges of the gaps of the free operator, for small random perturbations [281].

6.1.6 Mean Square Displacement

Let $r(t)$ be the mean distance traveled in a time t by a wave packet $\psi_0(x)$, that is, $r(t) = \|x\psi_t(x)\|$. For periodic potentials one has $r(t) \sim Dt$ for some constant D. We say that one has strict localization if $r(t)$ is a.s. bounded. It follows from the Rage theorem [295] that this property implies the strong localization property. It is known that strong localization implies that $\lim r(t)/t = 0$ [296] but it turns out that in most situations where strong localization has been proved, one also gets strict localization (and also the fact that the averaged value of $r(t)$ is bounded). But there are localized models for which $r(t)$ grows faster than t^δ for any $0 < \delta < 1$.

Another concept is the so-called "weak localization" property. This means that there exists some parameter s in the model for which one has the diffusive behavior $r(t) \sim D(s)\sqrt{(t)}$ for large t, but $\lim_{s \to \infty} D(s) = 0$. A different use of the term "weak localization" occurs when the reflected wave is two times stronger in the incident direction than in other directions; this is also called "coherent backscattering" [15–17]. It is not clear whether these notions have any connection with the concept of strong localization, and this question certainly deserves particular attention.

The list of related works and authors is too long to be given here, and hence we restrict ourselves to two "textbooks" [297, 298] on these topics and the papers cited above where an extensive bibliography can be found.

6.2 Defining Mass Operators for Random Systems

Adriaan Tip

In this section we make use of the underlying probability measure to give a Hilbert-space definition of the mass operator self-energy $\Sigma(z)$, $\text{Im } z \neq 0$, for a random system. We find that it is a bounded operator, even if the random potential V in $H = H_0 + V$ is not bounded.

6.2.1 Introduction

Consider a random time evolution of the type

$$\partial_t F_\omega(t) = -iH_\omega F_\omega(t), \tag{6.5}$$

where H_ω is a self-adjoint operator acting in a Hilbert space $\mathcal{H}$, which depends on the random variable ω (for details of random operators, see [297]). A standard example is the Schrödinger equation for a particle traveling through an assembly of randomly distributed scattering centers. Each such assembly is a state of a random system, i.e. a realization of a random process, and is labeled by $\omega \in \Omega$, the state space of the random process. The occurrence of a specific realization is controlled by the probability measure $P(d\omega) \geq 0$, $\int_\Omega P(d\omega) = 1$. If $\Delta \subset \Omega$ then the probability to find the system in Δ is $P(\Delta) = \int_\Delta P(d\omega)$. In general $P(d\omega)$ is an abstract object but a simple example is that of N points, homogeneously distributed in the volume V, in which case ω corresponds to $\mathbf{y} = \{\mathbf{y}_1, \mathbf{y}_2, \ldots, \mathbf{y}_N\}$ with $\mathbf{y}_j \in V$ and $P(d\mathbf{y}) = V^{-N} d\mathbf{y}_1 \ldots d\mathbf{y}_N$. An important limiting case ($V \to \infty$, $n = N/V$, fixed) is the Poisson distribution. For Poisson-distributed scatterers, we have

$$H_\omega = \mathbf{p}^2/2m + V_\omega(\mathbf{x}) = H_0 + V_\omega(\mathbf{x}),$$

$$V_\omega(\mathbf{x}) = \sum_j \varphi(\mathbf{x} - \mathbf{y}_j), \tag{6.6}$$

where $\omega \longleftrightarrow \mathbf{y} = \{\mathbf{y}_1, \mathbf{y}_2, ...\mathbf{y}_j...\}$, $\mathbf{y}_j \in \mathbb{R}^3$, an infinite set of points without accumulation points. The Poisson probability measure is defined through its action on $\langle f, \mathbf{y} \rangle = \sum_j f(\mathbf{y}_j)$:

$$\int P(\mathrm{d}\mathbf{y}) \exp(-\langle f, \mathbf{y} \rangle) = \exp\left(-n \int_{\mathbb{R}^3} \mathrm{d}\mathbf{x}\{1 - \exp[-f(\mathbf{x})]\}\right). \tag{6.7}$$

All the usual Poisson formulas follow from this relation.

Formally we can define the mass operator $\Sigma(z)$ by the expectation E or average $\langle \ \rangle$ (Im $z \neq 0$):

$$E(z - H_\omega)^{-1} = \langle (z - H_\omega)^{-1} \rangle$$

$$= \int P(\mathrm{d}\omega)(z - H_\omega)^{-1} = (z - H_0 - \Sigma(z))^{-1}. \tag{6.8}$$

The operator $\Sigma(z)$ contains important information about the averaged behavior of the system, such as its density of states, and therefore it is useful to find out how a more rigorous definition can be given.

6.2.2 An Operator Expression for the Mass Operator

Going back to $F_\omega(\mathbf{x}, t)$ above, it will be clear that it is square-integrable in $\mathbf{x}$ if this is the case at the initial time. But then $\int P(\mathrm{d}\omega) \int \mathrm{d}\mathbf{x} |F_\omega(\mathbf{x}, t)|^2$ is also finite and we can consider it to be an element of the Hilbert space $\mathcal{K}$ of square-integrable functions of ω that take their value in $\mathcal{H}$:

$$F \in \mathcal{K} = L^2(\Omega, P(\mathrm{d}\omega); \mathcal{H}) = \mathcal{H}_0 \otimes \mathcal{H}; \tag{6.9}$$

$$\mathcal{H}_0 = L^2(\Omega, P(\mathrm{d}\omega)). \tag{6.10}$$

In $\mathcal{K}$ we can define H by

$$(Hf)(\omega) = H_\omega f(\omega), f(\omega) \in \mathcal{H}. \tag{6.11}$$

Note that $f \in \mathcal{H}$, independent of ω, is also contained in $\mathcal{K}$, so $\mathcal{H}$ is a subspace of $\mathcal{K}$. But now we can introduce a projector E in $\mathcal{K}$ through

$$(Ef)(\omega) = \int P(\mathrm{d}\omega)f(\omega). \tag{6.12}$$

Thus E maps $\mathcal{K}$ onto $\mathcal{H}$. Clearly $E^2 = E = E^*$, so E is indeed a projection operator. Now consider $E(z - H)^{-1}E$. We have, with $R(z) = (z - H)^{-1}$ and $R_\omega(z) = (z - H_\omega)^{-1}$

$$[ER(z)Ef](\omega) = \int P(\mathrm{d}\omega)R_\omega(z)(Ef)(\omega)$$

$$= \int P(\mathrm{d}\omega)R_\omega(z)Ef = [z - H_0 - \Sigma(z)]^{-1}Ef, \tag{6.13}$$

where we have used (6.8) and the fact that $(Ef)(\omega)$ does not depend on ω.

However, we can interpret the above in a different way by making use of the Feshbach projection formula [301]. Thus if H is an operator and P a projector, and $Q = 1 - P$, then (with $H_P \equiv PHP$, $H_{PQ} \equiv PHQ$, etc.)

$$
\begin{aligned}
(z - H)^{-1} &= (z - H_Q)^{-1} + [P + (z - H_Q)^{-1} H_{QP}] \\
&\quad \times G_P(z)[P + H_{PQ}(z - H_Q)^{-1}],
\end{aligned}
\tag{6.14}
$$

$$
G_P(z) = [z - H_{PP} - H_{PQ}(z - H_Q)^{-1} H_{QP}]^{-1}.
\tag{6.15}
$$

Comparing the two, we now identify, taking $P = \mathrm{E}$ and $Q = \mathrm{F}$,

$$
\mathrm{E}(z - H)^{-1} \mathrm{E} = \langle R_\omega(z) \rangle = G_\mathrm{E}(z)
\tag{6.16}
$$

and

$$
\Sigma(z) = H_{\mathrm{EE}} - H_0 + H_{\mathrm{EF}}(z - H_\mathrm{F})^{-1} H_{\mathrm{FE}}.
\tag{6.17}
$$

In actual cases $H_{\mathrm{EE}} = \int P(d\omega) H_\omega$ can be a simple expression. For the Poisson example it equals $H_0 + n\langle \varphi \rangle$, where n is the density of the scatterers and $\langle \varphi \rangle = \int d\mathbf{x}\varphi(\mathbf{x})$ is the average potential. Also, $(H_0)_{\mathrm{EF}} = (H_0)_{\mathrm{FE}} = 0$, so $H_{\mathrm{EF}} = V_{\mathrm{EF}}$. In other cases the situation can be different. In the electromagnetic situation the electric permeability $\varepsilon_\omega(\mathbf{x})$ enters multiplicatively [299, 300] and this alters matters. The right-hand side of (6.17) is a well-defined object and we can now verify its existence as an operator on $\mathcal{H}$.

6.2.3 Existence of $\Sigma(z)$

Recall that a dense set in a Hilbert space has the whole space as its closure and that the intersection of two dense sets need not be dense at all. Note further that a bounded operator has the whole space as its domain. Since $\Sigma(z)$ is part of $\langle R_\omega(z) \rangle$, the first thing is to verify that its inverse exists. Indeed, the following proposition is true.

Proposition 1. $\langle R_\omega(z) \rangle$ *is a bounded and invertible operator with a densely defined inverse. Moreover, it is analytic in* $\mathbb{C} \backslash \mathbb{R}$.

Here, we shall not give the (easy) proof, which can be found in [63], as can the proofs of the statements below. It follows that the inverse $\langle R_\omega(z) \rangle^{-1}$ is defined on a dense set in $\mathcal{H}$. Formally $\Sigma(z) = \langle R_\omega(z) \rangle^{-1} - H_0$, but although H_0 is also defined on a dense set, their intersection need not have this property, so we are not yet through. From (6.12),

$$
\Sigma(z) = \mathrm{E}V + \mathrm{E}[V(z - H_\mathrm{F})^{-1}V].
\tag{6.18}
$$

The following proposition makes sense out of this.

Proposition 2. *Assume that* $\mathrm{E}V(\mathbf{x})$ *and* $\mathrm{E}V(\mathbf{x})^2$ *are both finite (this is true for the Poisson case if* $|\varphi(\mathbf{x})|$ *and* $\varphi(\mathbf{x})^2$ *are integrable over* $\mathbf{x}$*). Then* H_E, $\mathrm{E}V$ *and* $V\mathrm{E}$ *are bounded operators on* $\mathcal{K}$. *In fact* $\mathrm{E}V$ *is a bounded operator from* $\mathcal{K}$ *to* $\mathcal{H}$, *and* $V\mathrm{E}$ *is from* $\mathcal{H}$ *to* $\mathcal{K}$.

This implies that (6.13) is a bounded operator on $\mathcal{H}$. Finally, we can state the following proposition.

Proposition 3. *For* $z \in \mathbb{C} \backslash \mathbb{R}$,

$$\mathrm{E}R(z)\mathrm{E} = [\mathrm{E}R(z)]\mathrm{E} = [z - H_0 - \Sigma(z)]^{-1},$$
$$\Sigma(z) = \mathrm{E}V + \mathrm{E}[V(z - H_{\mathrm{F}})^{-1}V], \tag{6.19}$$

where $\Sigma(z)$ *is a bounded operator on* $\mathcal{H}$, *analytic in* $z \in \mathbb{C} \backslash \mathbb{R}$.

Thus we have achieved our goal. In particular it is interesting to note that V_ω is usually not a bounded operator but that $\Sigma(z)$, which in a certain sense is an average over V_ω, is bounded. Note further that nothing has been said about the existence of the limits as z approaches the real axis from above or below. In fact this is a much harder problem.

Depending on the further properties of the probability measure, the structure of $\Sigma(z)$ may simplify. Again taking the Schrödinger Poisson case as an example, its translation and rotation invariances make Σ a multiplication operator in momentum space, depending only on the absolute value p of the momentum operator $\mathbf{p}$: $\Sigma = \Sigma(p, z)$. In the Maxwell case things are more complicated since inverses of square roots appear, such as $\varepsilon_\omega(\mathbf{x})^{-1/2}$ [299].

7. Seismology

Michel Campillo, Ludovic Margerin and Keiiti Aki

In recent years, the attention of seismologists has turned to the physics of multiple scattering of elastic waves in the Earth. Analysis of the records of ground motion at the surface of the Earth has allowed imaging of the interior of the planet at different scales, in the framework of the classical theory of elastodynamics. Seismic data consist of time series of the 3D motions recorded when the surface of the Earth is hit by mechanical waves. Seismic signals cover a frequency band typically between 0.01 and 100 Hz. The sources in most cases produce pulses which are very short with respect to the travel time of the waves.

The development of piecewise homogeneous Earth models has been a fundamental step forward, with important implications for studies of the Earth's dynamics, the origin of the magnetic field and the quantification of earthquakes. Whatever the particular technique for imaging in use (reflection seismic prospecting, body wave tomography or surface wave tomography), the objective has always been the determination of the local local properties of an effective medium that can be related to a particular constituent or geological unit. However, it was realized early that such models cannot account for all characteristics of seismic signals. Two major observations lead us to consider the effect of scattering by the inhomogeneity of the materials: the strong apparent attenuation of the seismic signals and the duration of the seismograms, which greatly exceeds the travel times of the direct waves. These late arrivals are called the seismic coda.

7.1 Specific Consequences of Elasticity

The equation of linear isotropic elasticity for the three components of the displacement $u_i(t, x_1, x_2, x_3), i = 1, 2, 3$, is

$$\rho \frac{\partial^2 u_i}{\partial t^2} = \frac{\partial}{\partial x_i} \left(\lambda \frac{\partial u_j}{\partial x_j} \right) + \frac{\partial}{\partial x_j} \left(\mu \frac{\partial u_j}{\partial x_i} + \mu \frac{\partial u_i}{\partial x_j} \right), \tag{7.1}$$

in which $\lambda(x_1, x_2, x_3)$ and $\mu(x_1, x_2, x_3)$ are the Lamé parameters and $\rho(x_1, x_2, x_3)$ is the density. In a homogeneous medium this equation reduces to two wave equations for the potentials: the scalar compressional potential φ and the vectorial shear potential ψ. These potentials are associated with the

longitudinal P wave and the transversely polarized S waves. They propagate
with different velocities α and β respectively, and couple during a scattering
event. Unlike what happens in classical optics or acoustics, the coupling is
more complex than just a mixing of polarization channels, since two different
waves and wave speeds are involved. This aspect has an interesting analogy
in the recent study of light diffusion in nematic liquid crystals [79, 81, 82]
(Sect. 1.4.3).

In the context of elastic-wave scattering, it is important to note that
P to S and S to P conversions are not processes of equivalent strength. As
demonstrated from the reciprocity theorem, for a single scattering event [302]
(corrected in [303]), the scattering coefficients in terms of energy, g_{PS} and g_{SP},
are related by

$$\frac{g_{PS}}{g_{SP}} = 2\frac{\alpha^4}{\beta^4} \,. \tag{7.2}$$

This result indicates that at each diffraction event the energy is preferentially diffracted into shear waves. So the shear energy will finally dominate
the diffuse field. This property has been used as an argument to limit the
discussion on multiply scattered waves to the simplest case of shear waves.
The observations confirm the prominence of shear waves in the late arrivals.
Nevertheless, we shall show that a complete treatment of the elastic case is
required.

7.2 1D Problem

The one-dimensional problem has been extensively studied since the shallow structure of the Earth can often be considered as a stack of thin layers
with different densities and wave velocities that fluctuate randomly. This is
particularly true for sedimentary structures from which hydrocarbons are ex-
tracted. In the long-wavelength limit, such a medium can be described by the
effective anisotropic parameters [304]. For high-frequency waves, it has been
noticed in seismic prospecting that the frequency content of a pulse propagating through a thin, layered structure changes drastically, resulting in the
phenomenon known as "stratigraphic filtering". This effect is indeed of great
practical importance in the signal processing of reflected pulses from deep re-
flectors. It would also be a valuable parameter to measure, if one could relate
the apparent attenuation of the pulse to the statistical characteristics of the
stratigraphy itself. This apparent attenuation is indeed directly associated
with the localization length $\mathcal{L}$, although this terminology is not currently
used in the seismological literature. The spatial decay is commonly characterized by the scattering quality factor $Q(f)$, originating from a confusing
analogy with absorption attenuation. $Q(f)$ and $\mathcal{L}(f)$ are simply related by

$$\mathcal{L}(f) = \frac{Q(f)V_e}{\pi f}, \tag{7.3}$$

where V_e is the effective velocity and f the frequency.

7.2.1 Acoustic Waves

Multiple scattering in a layered structure was first studied for acoustic waves, corresponding to a scalar simplification in which the coupling between the two elastic potentials is neglected.

In 1971, O'Doherty and Anstey [305] proposed an expression relating the amplitude spectrum of the stratigraphic filter $A(\omega)$ for a vertically incident wave to the power spectrum of the reflection coefficient series $R(\omega)$ and the travel time t of the transmitted primary wave, in the form

$$A(\omega) = \exp[-R(\omega)t]. \tag{7.4}$$

This formula has been derived [306] in the context of the mean-field formalism. This method is based on the representation of the field by the sum of a mean field (in the sense of ensemble average) and a small fluctuating field. Weak fluctuations and a homogeneous background are assumed. The property of self-averaging is implicit in the use of this expression.

Asch et al. [307] give a comprehensive analysis of the properties of an acoustic signal reflected by a randomly layered medium that obeys the scaling property (see also Sect. 4.3)

$$a < \lambda < L, \tag{7.5}$$

where a is a characteristic length scale of the inhomogeneity (the thickness of the layers), λ the wavelength and L the thickness of the inhomogeneous slab. This regime involves a low-frequency approximation and holds for several experimental configurations. A review of the theory of localization for acoustic waves can be found in [217]. A remarkable application to seismic data was proposed by using an approximate expression for the localization length [308]. The following functional dependence of the localization length on the frequency has been proposed [309]:

$$\mathcal{L}(f) = C_1 + \frac{C_2}{f^2}, \tag{7.6}$$

where C_1 and C_2 are constants characterizing the medium; C_1 corresponds to the high-frequency limit, for which a lower bound of the extinction is given by taking into account only the product of all transmission coefficients of the primary pulse at every interface, which are independent of the frequency. The $1/f^2$ divergence of $\mathcal{L}(f)$ at low frequency can be intuitively explained by Rayleigh scattering. The important theoretical contribution was to show that C_2 can be computed from the correlation function of the fluctuations of compressibility [309]. As an illustration, for the case of nonstationary fluctuations of the medium characterized locally by an exponential correlation function, one obtains

$$C_2 = \frac{v_0^2}{2\pi^2} \frac{1}{\int_0^L c(z) \langle \sigma^2(z) \rangle \, dz}, \tag{7.7}$$

where $c(z)$ is the velocity correlation length of the small-scale inhomogeneity around the depth z and $\langle \sigma^2(z) \rangle$ is the relative variance of the compressibility. The second term in the right-hand side of (7.6) is interpreted as the contribution of multiple scattering in the forward direction. In this way it is possible to delineate almost explicitly the part played by the two competing physical process at work: single scattering and multiple forward scattering. Similarly, it was found that the spectrum of the backscattered field is directly related to the localization length. This simple formulation has been confirmed by numerical tests.

Time-domain properties of a propagating acoustic pulse have been studied under the scaling condition (7.5) in a layered medium with strong velocity perturbations, in 1D (the incident-plane-wave case) [310, 311]. It was demonstrated that propagation through the layered medium results in a time convolution of the incident pulse by a Gaussian distribution, whose variance increases with the travel time.

7.2.2 Elastic Waves

As was stated in Sect. 7.1, seismology actually deals with elastic waves rather than with scalar waves. A recent study [312] discusses a localization theory for fully elastic waves in a randomly layered medium. The evolution of the field can be described through two quantities. As for scalar waves, the localization length characterizes the spatial decay of the mean amplitude of the incident wave when propagating through the stack of random layers. The second quantity is the equilibration length, giving the characteristic length scale beyond which the ratio between the energies of compressional and shear waves tends to a constant. The existence of this equilibrium is a very important result, which will be discussed below. It is related to the nonsymmetric scattering coefficients from P to S and S to P waves, as expressed by (7.2).

The previous discussion concerned the restricted case of waves propagating in a stack of flat layers. This configuration was chosen to enable a rigorous mathematical treatment of the problem, at least under the scaling condition (7.5).

A similar configuration can be encountered in seismic prospecting in a sedimentary basin when waves reflected by a deep reflector are observed at the surface after having passed through a stack of thin layers, or when downgoing waves are recorded along a borehole in a vertical seismic-profile configuration. The signals generated by earthquakes are in most cases recorded at distances of several tens of kilometers, and therefore the wavefield propagates

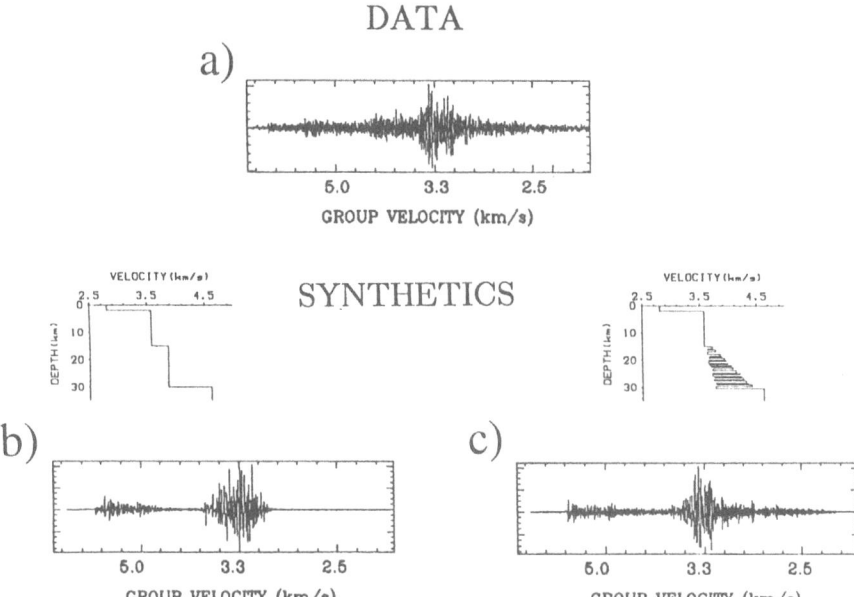

Fig. 7.1. (a) Example of a regional record of an earthquake at a distance of 386 km. The vertical ground velocity is plotted as a function of the group velocity. (b) Synthetic seismogram (*bottom*) computed for a three-layer crustal model (*top*). (c) Synthetic seismogram (*bottom*) computed for a crustal model with a thinly layered lower crust (*top*). The coda waves with low group velocities are associated with elastic leaky modes of the thin layering

in various types of geological structures in which the flat layering cannot always be regarded as a realistic approximation. Nevertheless, radiation from point sources in a 1D model has been successfully used for the study of surface waves or guided waves. As an example we present in Fig. 7.1a an actual seismogram recorded at regional distance. The records consist of the vertical ground-motion velocity. Because of the strong increase of wave velocity with depth in the shallow part of the Earth, the propagation of energy is at this distance almost horizontal, in the form of guided waves. The high-amplitude wave packet is a shear wave guided in the upper 30 km (the Earth's crust), which is called "Lg" in the seismological literature and has been extensively used in monitoring nuclear explosions. A complete numerical computation for a three-layer flat structure produces the right amplitude and time of onset of the wave train, in spite of the extreme oversimplification of the Earth structure (Fig. 7.1b). The main discrepancy between the observations and synthetics concerns the late part of the seismograms. The slowly decaying tail in the observation is the seismic coda. It is absent in the synthetics since it is produced by multiple scattering of elastic waves in the Earth. A series of flat layers with alternating high and low velocities is a crude representation of the laminations in the lower crust revealed by deep seismic soundings. The model used in Fig. 7.1c is derived from an analysis of the filtering effect of the lower crust on deep reflections [313]. It is important to notice the strong increase of signal duration due to the presence of the layering at the base of the waveguide. This effect does not exist for scalar waves and is not predicted by a normal-mode representation of the guided elastic waves [314]. This early coda is the result of the contribution from leaky modes resulting from the multiple conversions of the field when it is scattered in the thin layering. This phenomenon is a special property of elastic waves guided in a stack of flat layers.

7.3 Seismic Coda

The most spectacular evidence of seismic scattering in the Earth is the long duration of seismic signals for local earthquakes recorded in a frequency range between 1 and 10 Hz. This duration greatly exceeds the travel time of direct paths. The late arrivals build up the "seismic coda" and this could be referred to as the "deep coda" in the case of earthquake data, a term that indicates its origin in the deep layers. It has been shown by Aki and Chouet [315] that the decay with time of the coda envelope is a constant regional characteristic, independent of the precise location of the station or of the earthquake, or of the earthquake magnitude. According to [316] this property of stationarity is observed for lapse times larger than twice the travel time of direct shear waves t_S. This rule of thumb has been widely used in coda studies, but it cannot be applied to the case where the source–receiver distance is smaller than the characteristic scale of the scattering process, i.e. the mean free path. The analysis of the seismic field using an array of seismometers indicates that the coda waves consist of shear waves coming from almost all directions. Array analysis can also be used in a reciprocal configuration [317]. In this case an "array" of earthquakes at depth is observed at the Earth's surface. This technique makes it possible to analyze the take-off of the waves at the source. Its application showed that only the early coda is made of waves scattered in the vicinity of the recorder (i.e. close to the surface), while after two or three times t_S the coda waves leave the source in a wide variety of directions [318].

The first attempt to interpret the time decay was to consider the coda waves in the framework of single scattering. For the simplest case of a homogeneous background, the scalar approximation and a receiver close to the source, the energy density for a frequency f at a lapse time t can be written as [315]

$$E(f,t) = \frac{S_0(f)g_\pi(f)}{2\pi\beta^2 t^{2\gamma}} \exp\left(-\frac{2\pi f t}{Q_c}\right), \qquad (7.8)$$

where $S_0(f)$ is the shear energy emitted by the source, $g_\pi(f)$ is the backscattering coefficient and β is the shear wave velocity; γ is a factor of spreading depending upon the type of wave considered (0.5 for surface waves and 1 for body waves). Q_c, the "coda quality factor", expresses the energy decay due to scattering and inelastic processes. This expression has been widely used to characterize observations, and Q_c has emerged as a very stable parameter that seems to be correlated with the tectonic setting of the region where the measurement is done (see [319] for a review of early work on Q_c). The interest of seismologists in Q_c stems from the fact that except for Q_c, all measurements of the amplitude or attenuation of high-frequency (direct) waves are very difficult in the Earth's crust owing to focusing, defocusing and interference effects. The physical meaning of Q_c is nevertheless not clear, since the model is very crude (scalar waves in a homogeneous background) and the parts played by scattering and inelasticity are difficult to separate.

From a theoretical point of view, one expects the single-scattering model to be valid for short lapse times. The model has been refined by taking into account a finite source–receiver distance [320]. Finite-difference computations [321] have demonstrated the validity of the single-scattering model for weak heterogeneity in a 2D medium. Seismologists have made extensive use of the Born approximation, including refinement of the classical Chernov approach, for the evaluation of the extinction coefficient of transient signals in the presence of weak heterogeneity [322]. From these various studies, it emerges that for short-period seismic waves the scattering mean free path is in general of the order of tens of kilometers, a value that is intermediate between the wavelength (kilometers) and the total travel distance (a few hundreds of kilometers).

After considering the problems inherent in the phenomenological model of single scattering and the difficulties of computation in 3D with the wave equation, Wu used the stationary radiative transfer equation to obtain the total seismic energy [323]. He gave a solution for the evolution of the total energy with distance from the source for scalar waves in a homogeneous background.

These results have been used to separate the inelastic and scattering parts of the attenuation. Measurements in different regions of the world [324, 325] lead to very different figures for the ratio between intrinsic and scattering extinction lengths. In some cases these interpretations lead to very small values of the albedo which are almost in contradiction with the very existence of the coda.

An integral form of the transport equation can be introduced for isotropic scattering, in the form of a scattered-energy equation:

$$E(\mathbf{r}, t) = E_{\text{in}}\left(t - \frac{|\mathbf{r} - \mathbf{r_0}|}{\beta}\right) \frac{\exp(-\eta|\mathbf{r} - \mathbf{r_0}|)}{4\pi|\mathbf{r} - \mathbf{r_0}|^2}$$

$$+ \int_V \eta_s E\left(\mathbf{r_1}, t - \frac{|\mathbf{r_1} - \mathbf{r}|}{\beta}\right) \frac{\exp(-\eta|\mathbf{r_1} - \mathbf{r}|)}{4\pi|\mathbf{r_1} - \mathbf{r}|^2} \, d\mathbf{r_1}, \qquad (7.9)$$

where E_{in} is the incident energy and η_s is the scattering coefficient; η is the total attenuation coefficient, which includes scattering and absorption. This equation has been solved for high-order scattering [326]. This formulation made clear to seismologists the closeness between the ray-theory and energy-transport approaches.

Gusev and Abubakirov [327] were the first to apply a Monte Carlo simulation of the time-dependent radiative transfer equation to seismology. While limited to scalar waves in a homogeneous medium, their computation showed the transition from single scattering to multiple scattering, towards the diffusion regime. Hoshiba's Monte Carlo simulation confirms the theoretical results of Wu for the total energy [328].

7.3.1 Inhomogeneous Diffusion Models

So far, the background model has been supposed to be homogeneous. This is a serious limitation for applications to the Earth, since the average velocities are known to vary strongly with depth, resulting in strong reflections and guided propagation. The transition between the low-velocity crust and the high-velocity mantle occurs at a depth of 30 to 40 km beneath the continents, where most of the data have been collected. On the other hand, it is expected that the density of scatterers (the heterogeneity) decreases with depth. In order to study the role played by the velocity change, we solved the radiative transfer equation by the Monte Carlo technique, taking into account the reflection at the base of the crust [329]. Three models were presented. The computations correspond to isotropic scattering with a mean free path of 30 km and a diffusion constant of 35 km^2/s. The first case is a homogeneous half, space with a uniform distribution of scatterers. In the second model ("Layer 1"), the scatterer distribution is limited to a depth of 40 km in a homogeneous background. In the third model ("Layer 2"), a velocity jump is added at the same depth of 40 km to account for the S-wave velocity contrast between the crust and the mantle (3.5 km/s and 4.7 km/s respectively). The envelopes of the scattered energy at a distance of 100 km from a surface point source are presented in Fig. 7.2. In addition to the Monte Carlo results, the solutions of the diffusion equation corresponding to each model have been computed. These solutions (or their leading terms) are shown as smooth lines in Fig. 7.2. In every case, there is good agreement between the radiative-transfer-equation and diffusion solutions for large lapse times. The early part of the envelope is enlarged to show the onset of critically reflected waves in the model Layer 2. Later on, these waves tend to trap a part of the scattered energy in the crust. This is why the envelope has a much slower decay in the presence of a velocity jump (Layer 2) than in a homogeneous half-space (Layer 1). As shown in the figure, the solution of the diffusion equation provides a very good expression for the asymptotic behavior of the transfer equation. The functional forms in the diffusive regimes are the following, for $t \to \infty$:

$$\text{halfspace}: \quad E(t) \propto t^{-3/2} \tag{7.10}$$

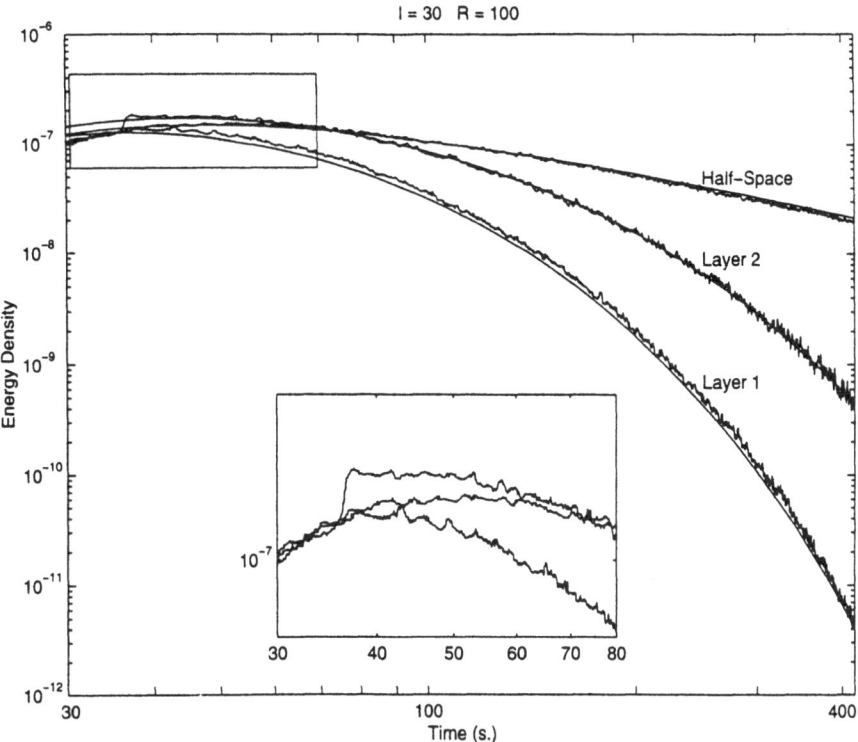

Fig. 7.2. Comparison of the results for a half-space and the models "Layer 1" and "Layer 2". The solutions of both the radiative transfer and the diffusion equation are shown. The former exhibits the ripples characteristic of the Monte Carlo method used. The source–station distance is $R = 100$ km and a mean free path $\ell^* = 30$ km has been adopted in all models. For Layer 1 and Layer 2, scattering is confined to a slab of thickness $H = 40$ km. In the half-space model and Layer 1, the S-wave velocity is uniform and equal to 3.5 km/s, whereas Layer 2 includes a velocity jump at depth H. The time origin is time of the release of energy at the source. The beginning of the signals is magnified (*inset*) in order to distinguish the evolution of the coda in the three models

$$\text{Layer 1 and Layer 2}: \quad E(t) \propto \frac{\exp(-Dt\xi_0^2/H^2)}{H\,D\,t}. \tag{7.11}$$

Here D is the diffusion coefficient and H is the thickness of the layer; ξ_0^2 is a variable depending upon H and D in a homogeneous background (model Layer 1), while it is also a function of the reflection coefficient in the presence of a velocity contrast (model Layer 2) [329]. The half-space model does not predict the observed behavior in (7.8), which indicates a faster decay of energy with time. In the case of the models Layer 1 and Layer 2, the expression for the asymptotes of the envelopes is formally similar to the one proposed to fit the observations (7.8) and therefore makes it possible to give a physical interpretation of Q_c in terms of long-range diffusion with bottom leaks. A

quantitative analysis with realistic values shows that the model Layer 1 leads
to too rapid a decay of the envelope with respect to the observations. When
the reflected waves are taken into account (model Layer 2), the decay is close
to that observed [329], even for the perfectly elastic model considered here.
Typically, if one interprets the curve shown in Fig. 7.2 for the model Layer 2
in terms of (7.8), the Q for the coda is found to be frequency-dependent in the
form $Q_c = 230f$, which is definitely in the range of proposed values in areas
of active tectonics. This suggests that the albedo of the Earth's crust is much
larger than the values proposed using models with a uniform distribution of
scatterers.

7.3.2 Role of Polarization

In the previous discussion it was assumed that coda waves consist of shear
waves and that P to S and S to P conversions can be neglected. This assumption was suggested by (7.2). More generally, the transport equation can
be modified into two coupled equations for the P and S energies by taking
into account the coupling using the single-scattering result of (7.2) [330, 331].
More detailed theoretical discussions that consider the coupled equations for
the P-wave energy and the coherence matrix of the transversely polarized
shear waves have been given by Weaver [332] and Ryjhik et al. [333].

For an elastic wave, we have to define the scattering mean free paths
associated with four different types of scattering: PP, SS, PS and SP. For
each case, the mean free path ℓ^{XY} is given by the usual definition,

$$\frac{1}{\ell^{XY}} = \int_{-1}^{1} \sigma^{XY}(\cos\theta)\, \mathrm{d}\cos\theta, \tag{7.12}$$

where θ is the angle between the incident and scattered waves and σ^{XY} is
the differential cross-section for scattering of type XY, XY being PP, SS, PS
or SP. The corresponding attenuation coefficient is given by $\Sigma^{XY} = 1/\ell^{XY}$.
The total scattering mean free paths or reciprocal attenuation coefficients
can be written as

$$\frac{1}{\ell^P} = \frac{1}{\ell^{PP}} + \frac{1}{\ell^{PS}}, \tag{7.13}$$

$$\frac{1}{\ell^S} = \frac{1}{\ell^{SS}} + \frac{1}{\ell^{SP}}. \tag{7.14}$$

Recalling the relation (1.13), a relevant quantity is the forward weighted
scattering attenuation coefficient [334]

$$\Sigma^{*XY} = \int_{-1}^{1} \sigma^{XY}(\cos\theta)\cos\theta\, \mathrm{d}\cos\theta. \tag{7.15}$$

This quantity vanishes if forward scattering and backscattering have equal
probability and is intimately related to the transport mean free path as defined in optics according to (1.13). Since for each wave type the transport

mean free path describes the effect of both diffraction and mode conversion, its value cannot be deduced from the mean free path and the mean cosine of a scattering angle as for scalar waves. When there is no preferential scattering (all $\Sigma^{*XY} = 0$), the transport mean free path is equal to the mean free path, as for scalar waves. In the general case of preferential scattering [332], two transport mean free paths can be defined as

$$
\ell^{P*} = \frac{\Sigma^S - \Sigma^{*SS} + \Sigma^{*PS}}{(\Sigma^P - \Sigma^{*PP})(\Sigma^S - \Sigma^{*SS}) - \Sigma^{*PS}\Sigma^{*SP}},
$$

$$
\ell^{S*} = \frac{\Sigma^P - \Sigma^{*PP} + \Sigma^{*SP}}{(\Sigma^P - \Sigma^{*PP})(\Sigma^S - \Sigma^{*SS}) - \Sigma^{*PS}\Sigma^{*SP}}.
\tag{7.16}
$$

A striking feature that appears to be a universal property has been pointed out for the diffusive regime [333, 334]. For long lapse times when the diffusion approximation is valid, the theory predicts that, independently of the details of the scattering, the energies of the P and S waves tend to equilibrate at a constant ratio:

$$
\frac{E^P}{E^S} = \frac{\beta^3}{2\alpha^3}.
\tag{7.17}
$$

This important property remains to be confirmed by observations; it is also predicted in nematic liquid crystals [31]. The transition to the diffusive regime has been studied using a Monte Carlo simulation for elastic waves in the framework of ultrasound propagation [334]. It appears that the equilibrium between compressional and shear energy is preceded by an intermediate stage in which the shear component tends to become isotropic. The diffusion coefficient for elastic waves, D_e, is given by

$$
D_e = \frac{\beta\ell^{S*}}{3} \left(\frac{\alpha\ell^{P*}/\beta\ell^{S*} + 2(\alpha/\beta)^3}{1 + 2(\alpha/\beta)^3} \right).
\tag{7.18}
$$

For actual applications in seismology, the consideration of elastic waves makes a significant difference from the case of the scalar approximation. With realistic values the elastic diffusivity is about 50 % higher than the shear-only diffusivity [334].

7.4 Coherent Backscattering and Localization in Seismology

Except for the case of sedimentary (1D) layering, localization of seismic waves has so far not been demonstrated theoretically. The coherent backscattering cone, often believed to be a precursor of strong localization, has not been observed. The coherent backscattering cone is a phenomenon resulting from an interference between reciprocal paths that occurs in the exact direction of the incident wave, with a perfect amplitude-doubling in the case where reciprocity applies perfectly. In an elastic body such as the Earth, the scattering process includes the coupling between the two different wave types described above. Therefore the scattered field is made up of contributions from both P and S waves. The reciprocity principle cannot be invoked for the part of the field which is not of the same type as the incident wave, even in the specular direction. This suggests that the magnitude of this effect will be smaller for elastic waves than for scalar waves. From a practical point of view, the controlled sources of high energy are explosions, whose primary radiation is essentially P-like, whereas the scattered waves are essentially S-like, as emphasized before. This particular fact makes it difficult to realize an experiment that allows a direct measurement of the amplitude of the cone.

Coherent backscattering of elastic waves from a rough boundary has been computed numerically by Schultz and Toksoz [335, 336]. To obtain stabilization of the speckle, it was necessary to perform a configuration average over a large number of realization of the interface geometry. Such an averaging is very difficult to realize with actual data since self-averaging is not expected in reflection.

The analysis of seismological data shows that narrow regions exist in which the coherent shear waves disappear almost completely, while there is no evidence or indication of a strong intrinsic attenuation. The existence of extreme heterogeneity has been invoked to explain these extinctions [337].

The coda waves recorded during local earthquakes on the volcanic island of La Réunion by an array of seismometers show an intriguing behavior that seems to indicate some form of localization [338]. The situation in La Réunion is exceptional, with the presence of a magmatic zone which is known to be extremely heterogeneous and to have a high average velocity. This last point excludes the existence of trapped modes inside the magmatic zone. The local amplification factor associated with each station was computed from the coda wave amplitude at a long lapse time (70 s). The frequency band considered was 1–3 Hz. These amplification factors have been shown to correspond to the response of shallow layers beneath the station [339], and do not exhibit any systematic geographic dependence. The travel time t_S of direct waves in this experiment depends on the earthquake and is typically less than 10 s. One would expect that normalization done using the amplification factor would be effective for lapse times larger than $2t_S$, according to numerous observations around the world. A major departure from this usual behavior has been observed in La Réunion. To visualize this, the amplitude of the coda waves in the time window 30–40 s was corrected using the amplification factor and normalized with respect to a reference station. The distribution of the normalized coda amplitude shows a maximum, and a smooth decrease around the maximum. For all earthquakes, the maximum corresponds to the axis of the volcano, while some of the earthquakes are located far away. From the spread of this maximum (about 5 km) at a lapse time of 35 s, a rough estimate of the diffusion coefficient can be proposed. With the simple assumption of scalar transport theory, the transport mean free path can be evaluated. With the values obtained in this study, ℓ^* is found to be very short: of the order of 360 m. This is significantly smaller than the wavelength (about 1 km) and by the Ioffe–Regel criterion (1.20), localization concepts may be necessary to explain the observations.

References

[1]   S. Chandrasekhar, *Radiative Transfer* (Oxford University Press, London, New York, Dover, New York, 1950; 1960).

[2]   H.C. van de Hulst, *Multiple Light Scattering*, Vols. I and II (Academic, New York, 1980).

[3]   P.W. Anderson, Phys. Rev. **109**, 1492 (1958).

[4]   J. Bardeen, L.N. Cooper and J.R. Schrieffer, Phys. Rev. **108**, 1175 (1957).

[5]   B.L. Altschuler, A.G. Aronov and B.Z. Spivak, JETP Lett. **33**, 94 (1981).

[6]   A. Kawabata, in: *Anderson Localization*, edited by Y. Nagaoka and H. Fukuyama (Springer, Berlin, Heidelberg, 1982).

[7]   G. Bergmann, Phys. Rep. **101**, 1 (1984).

[8]   D.Yu. Sharvin and Yu.V. Sharvin, JETP Lett. **34**, 272 (1981).

[9]   H. Vloerbergsh, C. van Haesendonc, Y. Bruynseraede, A.H. Verbruggen, P.A.M. Holweg and S. Radelaar, Physica **175**, 217 (1991).

[10] R.A. Webb, S. Washburn, C.P. Umbach, R.B. Laibowitz, Phys. Rev. Lett. **54**, 2696 (1985).

[11] K. von Klitzing, G. Dorda and M. Pepper, Phys. Rev. Lett. **45**, 494 (1980).

[12] P.A. Lee and A.D. Stone, Phys. Rev. Lett. **55**, 1622 (1986).

[13] P.A. Lee, Physica A **140**, 169 (1986).

[14] P.W. Anderson, Phil. Mag. B **52**, 505 (1985).

[15] M.P. van Albada and A. Lagendijk, Phys. Rev. Lett. **55**, 2692 (1985).

[16] P.E. Wolf and G. Maret, Phys. Rev. Lett. **55**, 2696 (1985).

[17] M. Kaveh, M. Rosenbluh, I. Edrei and I. Freund, Phys. Rev. Lett. **57**, 2049 (1986).

[18] Y. Kuga and A. Ishimaru, J. Opt. Soc. Am. A **8**, 831 (1984).

[19] D.S. Wiersma, M.P. van Albada and A. Lagendijk, Phys. Rev. Lett. **75**, 1739 (1995); D.S. Wiersma and A. Lagendijk, Physics World, January 1997, page 33; D.S. Wiersma and A. Lagendijk, Phys. Rev. E **54**, 4256 (1996).

[20] E. Akkermans, P.E. Wolf and R. Maynard, Phys. Rev. Lett. **56**, 1471 (1986).

[21] F.A. Erbacher, R. Lenke and G. Maret, Europhys. Lett. **21**, 55 (1993).

[22] A.S. Martinez and R. Maynard, Phys. Rev. B. **50**, 3714 (1994).

[23] B.A. van Tiggelen, R. Maynard and T.M. Nieuwenhuizen, Phys. Rev. E **53**, 2881 (1996).

[24] I. Freund, M. Rosenbluh, R. Berkovits and M. Kaveh, Phys. Rev. Lett. **61**, 1214 (1988).

[25] V.M. Agranovich and V.E. Kravtsov, Phys. Rev. B **43**, 13691 (1991).

[26] A. Heiderich, R. Maynard and B.A. van Tiggelen, Opt. Commun. **115**, 392 (1995).

[27] A. Heiderich, A.S. Martinez, R. Maynard and B.A. van Tiggelen, Phys. Lett. A **185**, 110 (1994).

[28] A.H. Gandjbakhche, R.F. Bonner and R. Nossal, J. Stat. Phys. **69**, 35 (1992).

[29] G. Maret and P.E. Wolf, Z. Phys. B **65**, 409 (1987).

[30] G. Maret, Current Opinion in Colloid & Interface Science **2**, 251 (1997).

[31] H.K.M. Vithana, L. Asfaw and D.L. Johnson, Phys. Rev. Lett. **70**, 3561 (1993).

[32] G. Bayer and T. Niederdränk, Phys. Rev. Lett. **70**, 3884 (1993).

[33] N. Garcia and A.Z. Genack, Phys. Rev. Lett. **66**, 1850 (1991).

[34] R. Deliaouch, J.P. Armstrong, S. Schultz, P.M. Platzman and S.L. McCall, Nature **354**, 53 (1991).

[35] B.A. van Tiggelen and R. Maynard, in: *Wave Propagation in Complex Media*, edited by G. Papanicolaou (Springer, Berlin, Heidelberg, 1997).

[36] D.S. Wiersma, M.P. van Albada, B.A. van Tiggelen and A. Lagendijk, Phys. Rev. Lett. **74**, 4193 (1995). B.A. van Tiggelen, D.S. Wiersma and A. Lagendijk, Europhys. Lett. **30**, 1 (1995).

[37] B.A. van Tiggelen, Phys. Rev. Lett. **75**, 422 (1995).

[38] G.L.J.A. Rikken and B.A. van Tiggelen, Nature **381**, 54 (1996).

[39] D.J. Bicout, E. Akkermans and R. Maynard, J. Phys. I (France) **1**, 471 (1991).

[40] D.J. Bicout and G. Maret, Physica A **210**, 87 (1994).

[41] D.J. Bicout and R. Maynard, Physica B **204**, 20 (1995).

[42] M. Heckmeier and G. Maret, Europhys. Lett. **34**, 257 (1996).

[43] A.Z. Genack, W. Polkosnick, A.A. Lisyanski, N. Garcia and P. Sebbah, in: *OSA Proceedings on Advances in Optical Imaging and Photon Migration*, edited by R.R. Alfano (Optical Society of America, Washington, DC, 1994), p. 43.

[44] G. Jarry, E. Steimer, V. Damaschini, M. Jurczak and R. Kaiser, J. Opt. **28**, 83 (1997).

[45] M. Kempe, A.Z. Genack, W. Rudolph and P. Dorn, J. Opt. Soc. Am. A **14** (1), 216 (1997).

[46] B.A. van Tiggelen and A. Lagendijk, Europhys. Lett. **23**, 311 (1993).

[47] A. Lagendijk and B.A. van Tiggelen, Phys. Rep. **270**, 143 (1996).

[48] M.P. van Albada, B.A. van Tiggelen, A. Lagendijk and A. Tip, Phys. Rev. Lett. **66**, 3132 (1991).

[49] Yu.N. Barabanenkov and V.D. Ozrin, Phys. Rev. Lett. **69**, 1364 (1992).

[50] B.A. van Tiggelen, A. Lagendijk and A. Tip, Phys. Rev. Lett. **71**, 1284 (1993); Yu.N. Barabanenkov and V.D. Ozrin, Phys. Rev. Lett. **71**, 1285 (1993).

[51] E. Kogan and M. Kaveh, Phys. Rev. B **46**, 10636 (1992).

[52] G. Cwilich and Y. Fu, Phys. Rev. B **46**, 12015 (1992).

[53] M. Kafesaki and E.N. Economou, Europhys. Lett. **37**, 7 (1997).

[54] K. Busch, C.M. Soukoulis and E.N. Economou, Phys. Rev. B **52**, 10834 (1995).

[55] K. Busch and C.M. Soukoulis, Phys. Rev. Lett. **75**, 3442 (1995).

[56] J. Kroha, C.M. Soukoulis and P. Wölfle, Phys. Rev. B **47**, 9208 (1992).

[57] D. Livdan and A.A. Lisyansky, Phys. Rev. B **53**, 14843 (1996).

[58] Yu.N. Barabanenkov, L.M. Zurk, M.Yu. Barabanenkov, J. Electromagnetic Waves and Applications (JEWA) **11**, 293. (1997); ibid. **9**, 1393 (1995).

[59] Y. Kuga, A. Ishimaru and D. Rice, Phys. Rev. B **48**, 13155 (1993).

[60] P. Sheng, M. Zhou and Z.Q. Zhang, Phys. Rev. Lett. **72**, 234 (1994).

[61] M. Stoytchev, N. Garcia and A.Z. Genack, in: *OSA TOPS on Advances in Optical Imaging and Photon Migration 1996*, edited by R.R. Alfano and J.G. Fujimoto (Optical Society of America, Washington, DC, 1996), p. 383.

[62] U. Frisch, *Probabilistic Methods in Applied Mathematics*, Vols. I and II, edited by A.T. Bharucha-Reid (Academic, New York, 1968).

[63] A. Tip, J. Math. Phys. **35**, 113 (1994).

[64] M. Born and E. Wolf, *Principles of Optics* (Pergamon, Oxford, 1975).

[65] H.A. Lorentz, Wiedem. Ann. **9**, 4641 (1880).

[66] B.U. Felderhof, G.W. Ford and E.G.D. Cohen, J. Stat. Phys. **33**, 1614 (1983).
[67] A. Lagendijk, B. Nienhuis, B.A. van Tiggelen and P. de Vries, Phys. Rev. Lett. **79**, 657 (1997).
[68] V. Twersky, J. Math. Phys. **18**, 2468 (1977).
[69] V.A. Davis and L. Schwarz, Phys. Rev. B **31**, 5155 (1985).
[70] C.G.B. Garret and D.E. McCumber, Phys. Rev. A **1**, 305 (1970).
[71] S. Chu and S. Wong, Phys. Rev. Lett. **48**, 738 (1982).
[72] G.C. Papanicolaou and R. Burridge, J. Math. Phys. **16**, 2074 (1975).
[73] J.W. Goodman, *Statistical Optics* (Wiley, New York, 1985).
[74] D.J. Pine, D.A. Weitz, G. Maret, P.E. Wolf, E. Herbolzheimer and P.M. Chaikin, in: *Scattering and Localization of Classical Waves in Random Media* edited by P. Sheng (World Scientific, Singapore, 1990).
[75] J.F. de Haan, P.B. Bosma and J. Hovenier, Astr. Astrophys. **183**, 371 (1987).
[76] A. Ishimaru, *Wave Propagation in Random Media*, Vols. 1 and 2 (Academic, New York, 1978).
[77] J.H. Li, A.A. Lisyanski, T.D. Cheung, D. Livdan and A.Z. Genack, Europhys. Lett. **22**, 675 (1993).
[78] J.M. Tualle, B. Gelebart, E. Tinet, S. Avrillier and J.P. Ollivier, Opt. Commun. **124**, 216 (1996).
[79] B.A. van Tiggelen, R. Maynard and A. Heiderich, Phys. Rev. Lett. **77**, 639 (1996).
[80] A. Heiderich, R. Maynard and B.A. van Tiggelen, J. Phys. II (France) **7**, 765 (1997).
[81] H. Stark and T.C. Lubensky, Phys. Rev. Lett. **77**, 2229 (1996).
[82] M.H. Kao, K.A. Jester, A.G. Yodh and P.J. Collins, Phys. Rev. Lett. **77**, 2233 (1996).
[83] B.A. van Tiggelen, R. Maynard and A. Heiderich, Mol. Liq. and Liq. Cryst. **293**, 205 (1997).
[84] H. Stark and T.C. Lubensky, Phys. Rev. E **55**, 514 (1997).
[85] D.Y. Ivanov and A.F. Kostko, Opt. Spectrosk. (USSR) **55**(5), 950 (1983).
[86] A.A. Golubentsev, Sov. Phys. JETP **59**, 26 (1984).
[87] D.J. Pine, D.A. Weitz, P.M. Chaikin and E. Herbolzheimer, Phys. Rev. Lett. **60**, 1134 (1988).
[88] D.A. Weitz and D.J. Pine, in: *Dynamic Light Scattering*, edited by W. Brown (Oxford University Press, New York, 1993), Chap. 16, pp. 652–720.
[89] D.A. Boas, L.E. Campbell and A.G. Yodh, Phys. Rev. Lett. **75**, 1855 (1995).
[90] K. Schätzel, M. Drewel and J. Ahrens, J. Phys. Condens. Matter **2**, SA393 (1990).
[91] H. Wiese and D. Horn, J. Chem. Phys. **94**, 6429 (1991).
[92] H.S. Dhadwal, R.R. Ansari and W.V. Meyer, Rev. Sci. Instrum. **62**, 2963 (1991).
[93] E.R. Van Keuren, H. Wiese and D. Horn, Coll. Surf. A: **77**, 29 (1993).
[94] D.A. Weitz, D.J. Pine, P.N. Pusey and R.J.A. Tough, Phys. Rev. Lett. **63**, 1747 (1989).
[95] J.X. Zhu, D.J. Durian, J. Müller, D.A. Weitz and D.J. Pine, Phys. Rev. Lett. **68**, 2559 (1992).
[96] M.H. Kao, A.G. Yodh and D.J. Pine, Phys. Rev. Lett. **70**, 242 (1993).
[97] X.L. Wu, D.J. Pine, P.M. Chaikin, J.S. Huang and D.A. Weitz, J. Opt. Soc. Am. B **7**, 15 (1990).
[98] D. Bicout, E. Akkermans and R. Maynard, J. Phys. (France) I **1**, 471 (1991).
[99] D. Bicout and G. Maret, Physica A **210**, 87 (1994).
[100] D. Bicout and R. Maynard, Physica B **204**, 20 (1995).
[101] J.M. Ginder, Phys. Rev. E **47**, 3418 (1993).

[102] W. Leutz and G. Maret, Physica B **204**, 14 (1995).
[103] D.J. Durian, D.A. Weitz and D.J. Pine, Science **252**, 686 (1991).
[104] D.J. Durian, D.A. Weitz and D.J. Pine, Phys. Rev. A **44**, 7902 (1991).
[105] A.G. Yodh and B. Chance, Physics Today **48**, 34 (1995).
[106] P.N. DenOuter, T.M. Nieuwenhuizen and A. Lagendijk, J. Opt. Soc. Am. A **10**, 1209 (1993).
[107] R. Berkovits and S. Feng, Phys. Rev. Lett. **65**, 3120 (1990).
[108] D.A. Boas and A.G. Yodh, J. Opt. Soc. Am. A **14**, 192 (1997).
[109] M. Heckmeier and G. Maret, Europhys. Lett. **34**, 257 (1996).
[110] M. Heckmeier, S.E. Skipetrov, G. Maret and R. Maynard, J. Opt. Soc. Am. A**14**, 185 (1997).
[111] M. Heckmeier and G. Maret, J. Coll. Int. Sci., in press (1997).
[112] M. Heckmeier and G. Maret, preprint.
[113] M.B. van der Mark, M.P. van Albada and A. Lagendijk, Phys. Rev. B **37**, 3575 (1988).
[114] E. Amic, J.M. Luck and T.M. Nieuwenhuizen, J. Phys. A: Math. Gen. **29**, 4915 (1996).
[115] E. Amic, J.M. Luck and T.M. Nieuwenhuizen, J. Phys. I (France) **7**, 445 (1997).
[116] M.P. van Albada, M.B. van der Mark and A. Lagendijk, J. Phys. D: Appl. Phys. **21**, S28 (1988).
[117] M.J. Stephen and G. Cwilich, Phys. Rev. B **34**, 7564 (1986).
[118] K.J. Peters, Phys. Rev. B **46**, 801 (1992).
[119] M.P. van Albada and A. Lagendijk, Phys. Rev. B **36**, 2353.
[120] A. Lagendijk, R. Vreeker and P. de Vries, Phys. Lett. A **136**, 81 (1989).
[121] T.M. Nieuwenhuizen and J.M. Luck, Phys. Rev. E **48**, 569 (1993).
[122] D.S. Wiersma, M.P. van Albada and A. Lagendijk, Rev. Sci. Instr. **66**, 5473 (1995).
[123] C. Goudeard, D. Husson, C. Sauteret, F. Auzel and A. Migus, J. Opt. Soc. Am. B **10**, 2358 (1993).
[124] N.M. Lawandy, R.M. Balachandran, A.S.L. Gomes and E. Sauvain, Nature **368**, 436 (1994).
[125] S. John and G. Pang, Phys. Rev. A **54**, 3642 (1996).
[126] B.A. van Tiggelen, D.S. Wiersma and A. Lagendijk, Europhys. Lett. **30**, 1 (1995).
[127] D. Vollhardt and P. Wölfle, Phys. Rev. B. **22**, 4666 (1980).
[128] E. Abrahams, P.W. Anderson, D.C. Licciardello and T.V. Ramakrishnan, Phys. Rev. Lett. **42**, 673 (1979).
[129] D. Vollhardt and P. Wölfle, "Self-Consistent Theory of Anderson Localization", in: *Electronic Phase Transitions*, edited by W. Hanke and Yu.V. Kopaev (North-Holland, Amsterdam, 1992).
[130] E. Hoffstetter and M. Schreiber, Phys. Rev. Lett. **73**, 3137 (1994).
[131] K.B. Efetov and A.I. Larkin, Sov. Phys. JETP **58**, 444 (1983).
[132] A. Sparenberg, G.L.J.A. Rikken and B.A. van Tiggelen, Phys. Rev. Lett. **79**, 757 (1997).
[133] P. Sheng and Z.Q. Zhang, Phys. Rev. Lett. **57**, 1897 (1986).
[134] C.M. Soukoulis, S. Datta and E.N. Economou, Phys. Rev. B **49**, 3800 (1994).
[135] G.L.J.A. Rikken and Y.A.R.R. Kessener, Phys. Rev. Lett. **74**, 880 (1995).
[136] G. Juzeliunas, Phys. Rev. A **55**, R4015 (1997).
[137] B.A. van Tiggelen and A. Lagendijk, Phys. Rev. B **50**, 16732 (1994).
[138] O. Morice, Y. Castin and J. Dalibard, Phys. Rev. A **51**, 3896 (1995).

[139] G.H. Watson, P.M. Saulnier, I.I. Tarhan and M.P. Zinkin, in: *Photonic Bandgaps and Localization*, edited by C.M. Soukoulis (Plenum, New York, 1993), p. 131.
[140] S. Fraden and G. Maret. Phys. Rev. Lett. **65**, 512 (1990).
[141] R. Berkovits and S. Feng, Phys. Rep. **238**, 135 (1994); C.W.J. Beenakker, Rev. Mod. Phys. **69**(3) 731 (1997).
[142] S. Feng, C. Kane, P.A. Lee and D.A. Stone, Phys. Rev. Lett. **61**, 834 (1988).
[143] A.Z. Genack, in: *Scattering and Localization of Classical Waves in Random Media*, edited by Ping Sheng (World Scientific, Singapore, 1990), p. 207–311.
[144] I. Freund, M. Rosenbluh and S. Feng, Phys. Rev. Lett. **61**, 2328 (1988).
[145] A.Z. Genack, N. Garcia and W. Polkosnik, Phys. Rev. Lett. **65**, 2129 (1990).
[146] J.F. de Boer, M.P. van Albada and A. Lagendijk, Phys. Rev. B **45**, 658 (1992).
[147] C.P. Umbach, S. Wasburn, R.B. Laibowitz and R.A. Webb, Phys. Rev. B **30**, 4048 (1984).
[148] E. Kogan, M. Kaveh, R. Baumgartner and R. Berkovits, Phys. Rev. B. **48**, 9404 (1993).
[149] N. Garcia and A.Z. Genack, Phys. Rev. Lett. **63**, 1678 (1993).
[150] T.M. Nieuwenhuizen and M.C.W. van Rossum, Phys. Rev. Lett. **74**, 2674 (1995).
[151] E. Kogan and M. Kaveh, Phys. Rev. B **52**, R3813 (1995).
[152] S.A. van Langen, P.W. Brouwer and C.W.J. Beenakker, Phys. Rev. E **53**, 1344 (1996).
[153] M. Stoytchev and A.Z. Genack, Phys. Rev. Lett. **79**, 309 (1997).
[154] D. Sornette and B. Souillard, Europhys. Lett. **13**, 7 (1996).
[155] D.T. Smithey, M. Beck and M.G. Raymer, Phys. Rev. Lett. **70**, 1244 (1993).
[156] M.G. Raymer, M. Beck and D.F. McAlister, Phys. Rev. Lett. **72**, 1137 (1994).
[157] D.F. McAlister, M. Beck, L. Clarke, A. Mayer and M.G. Raymer, Opt. Lett. **20**, 1181 (1995).
[158] R. Pnini and B. Shapiro, Phys. Lett. **157**, 265 (1991).
[159] P.A. Mello, E. Akkermans and B. Shapiro, Phys. Rev. Lett. **61**, 459 (1988).
[160] S. Feng, C. Kane, P.A. Lee and A.D. Stone, Phys. Rev. Lett. **61**, 839 (1988).
[161] B. Shapiro, Phys. Rev. Lett. **57**, 2168 (1986).
[162] N. Garcia, A.Z. Genack, R. Pnini and B. Shapiro, Phys.Lett.A **176**, 458 (1993).
[163] P. Sebbah, O. Legrand, B.A. van Tiggelen and A.Z. Genack, Phys. Rev. E. **56**, 3619 (1997).
[164] P. Sebbah, O. Legrand and A.Z. Genack, "Phase Statistics in Random Media", *Advances in Optical Imaging and Photon Migration – Orlando 96*, edited by R.R. Alfano and J.G. Fujimoto, Vol. 2, p. 386, (Optical Society of America, Washington, DC, 1996).
[165] A.Z. Genack, J.H. Li, N. Garcia and A.A. Lisyansky, in: *Photonic Band Gaps and Localization*, edited by C.M. Soukoulis (Plenum, New York, 1993), p. 23.
[166] B.A. van Tiggelen and E. Kogan, Phys. Rev. A **49**, 708 (1994).
[167] V. Gasparian and M. Pollack, Phys. Rev. B **47**, 2038 (1993).
[168] G. Iannaconne, Phys. Rev. B **51**, 4727 (1995).
[169] A. Kienle, L. Lilge, M.S. Patterson, R. Hibst, R. Steiner B.C. Wilson, Appl. Opt. **35**, 2304–2313 (1996).
[170] F. Bevilacqua, P. Marquet, C. Depeursinge and E. de Haller, Opt. Eng. **34**, 2064–2069 (1995).
[171] A.J. Welch and M.J.C. van Gemert, *Optical–Thermal Response of Laser-irradiated Tissue* (Plenum, New York, 1995).
[172] L. Wang and S.L. Jacques, SPIE **1888**, 107 (1993).
[173] P. Corcuff and J.L. Lévêque, Dermatology **186**, 50 (1993).

[174] M. Rajadhyaksha, M. Grossman, D. Esterowitz, R.H. Webb and R.R. Anderson, J. Invest. Dermatol. **104**, 946 (1995).

[175] B.R. Masters, G. Gonnord and P. Corcuff, J. Microsc. **185**, 329–338 (1997).

[176] J.M. Schmitt, A. Knüttel and M. Yadlovsky, J. Opt. Soc. Am. A **11**, 2226–2235 (1994).

[177] J.M. Schmitt, K. Ben-Letaief, J. Opt. Soc. Am. A **13**, 952–961 (1996).

[178] C.A. Puliafito, M.R. Hee, J.S. Schuman, J.G. Fujimoto, *Optical Coherence Tomography of Ocular Diseases*, (Slack Inc., 1996).

[179] S.A. Boppart, M.E. Brezinski, B. Bouma, G.J. Tearney and J.G. Fujimoto, Dev. Biol. **177**, 54 (1996).

[180] M.E. Brezinski, G.J. Tearney, B.E. Bouma, J.A. Izatt, M.R. Hee, E.A. Swanson, J.F. Southern, J.G. Fujimoto, Circulation **93**, 1206–1213 (1996).

[181] J.A. Izatt, M.R. Hee, G.M. Owen, E.A. Swanson and J.G. Fujimoto, Opt. Lett. **19**, 590 (1994).

[182] M. Kempe and W. Rudolph, Opt. Lett. **19**, 1919 (1994).

[183] M. Kempe, A. Thon and W. Rudolph, Opt. Commun. **110**, 492 (1994).

[184] T. Hellmuth, SPIE **2926**, 228–237 (1996).

[185] S. Fantini, M.A. Franceschini, J.S. Maier, S.A. Walker and B. Barbieri, E. Gratton, Opt. Eng. **34**, 32–42 (1995).

[186] I. Yoshiya, Y. Shimada and K. Tanaka, Med. Biol. Eng. Comput. **18**, 27–32 (1980).

[187] A. Knüttel, J.M. Schmitt and J.R. Knutson, Appl. Opt. **32**, 381 (1993).

[188] J.L. Bussière, J.C. Jouanin, E. Tinet, F. Revel, F. Bernard, S. Avrillier and J.P. Ollivier, Colloque OPT-DIAG 97, Paris (1997), Abstracts, p. 10.

[189] M. Anidjar, D. Ettori, O. Cussenot, P. Meria, F. Desgrandchamps, A. Cortesse, P. Teillac, A. Le Duc and S. Avrillier, J. Urology **156**, 1590–1596 (1996).

[190] P. Jichlinski, G. Wagnières, H.J. Leisinger and H. van den Bergh, Colloque OPT-DIAG 97, Paris (1997), Abstracts, p. 20.

[191] S. Delysse, J.M. Nunzi and P.L. Baldeck, Colloque OPT-DIAG 97, Paris (1997), Abstracts, p. 34.

[192] L. Wang, S.L. Jacques and X. Zhao, Opt. Lett. **20**, 629 (1995).

[193] M. Kempe, M. Larionov, D. Zaslavsky, A.Z. Genack, TOPS **2**, 328–331 (1996).

[194] L. Wang, X. Zhao and S.L. Jacques, TOPS **2**, 325–327 (1996).

[195] S. Lévêque, P. Gleyzes, C. Boccara, M. Lebec, M. Blanchot and H. Saint-Jalmes, Colloque OPT-DIAG 97, Paris (1997), Abstracts, p. 45.

[196] L. Garvican and P. Littlejohns, J. Med. Screening **3**, 123 (1996).

[197] R.T. Osteen, B. Cady, M. Friedman, W. Kraybill, S. Doggett, D. Hussey, M. Urist, J. Chmiel, R. Clive and D. Winchester, J. Nat. Cancer Inst. Monographs **16** (1994).

[198] R.J. Grable, SPIE **2979**, Paper 05 (1997).

[199] H. Jess, H. Erdl, K.T. Moesta, S. Fantini, M.A. Franceschini, E. Gratton and M. Kaschke, TOPS **2**, 126 (1996).

[200] J.H. Hoogenraad, M.B. van der Mark, S.B. Colak, G.W. 't Hooft, E.S. van der Linden, SPIE **3114**, in press.

[201] D.A. Beranon, J.P. van Houten, W.F. Cheong, E.L. Kermit and R.A. King, SPIE **2389**, 582 (1995).

[202] S.P. Gopinath, C.S. Robertson, R.G. Grossmann and B. Chance, J. Neurosurg. **79**, 43 (1993).

[203] A. Maki, Y. Yamashita, Y. Ito, E. Watanabe, Y. Manayagi and H. Koizumi, Med. Phys. **22**, 1997 (1995).

[204] A. Maki, Y. Yamashita, Y. Ito, E. Watanabe and H. Koizumi, TOPS **2**, 357–362 (1996).

[205] C.J.F. Böttcher, *Theory of Electric Polarization, Vol. I: Dielectrics in Static Fields* (Elsevier, Amsterdam, 1973).

[206] H.C. van de Hulst, *Light Scattering by Small Particles* (Dover, New York, 1981).

[207] J. Virmont and G. Ledanois, TOPS **2**, 307 (1996).

[208] D.A. Boas, M.A. O'Leary, B. Chance and A.G. Yodh, Appl. Opt. **36**, 75 (1997).

[209] E.T. Jaynes, *Papers on Probability and Statistical Physics*, edited by R.D. Rozencrantz (Reidel, Dordrecht, 1983).

[210] H. Heusmann, J. Kölzer and G. Mitic, J. Biomed. Opt. **1**, 425–434 (1996).

[211] S. Nioka and B. Chance, SPIE **3194**, (1997).

[212] K. Busch, C.M. Soukoulis and E.N. Economou, Phys. Rev. B **50**, 93 (1995).

[213] X. Jing, P. Sheng and M. Zhou, Phys. Rev. A **46**, 6513 (1992).

[214] P. Sheng, X. Jing and M. Zhou, Physica A **207**, 37 (1994).

[215] K. Busch and C.M. Soukoulis, Phys. Rev. B **54**, 893 (1996).

[216] A. Kirchner, K. Busch and C.M. Soukoulis, Phys. Rev. B **57**, 277 (1998).

[217] P. Sheng, *Introduction to Wave Scattering, Localization and Mesoscopic Phenomena*, (Academic Press, San Diego, 1995).

[218] N. Garcia, A.Z. Genack and A.A. Lisyansky, Phys. Rev. B **46**, 14475 (1992).

[219] E.N. Economou, C.M. Soukoulis and A.D. Zdetsis, Phys. Rev. B **30**, 1686 (1984).

[220] M.M. Sigalas, C.M. Soukoulis, C.T. Chan and D. Turner, Phys. Rev. B **53**, 8340 (1996).

[221] M. Fink, Physics Today, March 1997, p. 34.

[222] R. Mallart and M. Fink, J. Acoust. Soc. Am. **96**, 3721 (1994).

[223] A. Derode and M. Fink, J. Acoust. Soc. Am. **101**, 690 (1997).

[224] A. Tourin, P. Roux, A. Derode, B.A. van Tiggelen and M. Fink, Phys. Rev. Lett. **79**, 3637 (1997).

[225] M. Fink, Contemp. Phys. **37**, 95 (1996).

[226] A. Derode, P. Roux and M. Fink, Phys. Rev. Lett. **75**, 4206 (1995).

[227] A. Sommerfeld, Ann. Phys. (Leipzig) **44**, 177 (1914).

[228] L. Brillouin, *Wave Propagation in Periodic Structures* (Dover, New York, 1953); *Wave Propagation and Group Velocity* (Academic, New York, 1960).

[229] J. Page, P. Sheng, H.P. Schriemer, I. Jones, X. Jing and D. Weitz, Science **271**, 634 (1996).

[230] P. Lewicki, R. Burridge and G. Papanicolaou, Wave Motion **20**, 177–195 (1994).

[231] J.F. Clouet and J.P. Fouque, Ann. Appl. Probabil. **4**, 1083–1097 (1994).

[232] J. Chillan and J.P. Fouque, to appear in the SIAM J. Appl. Math. (1998).

[233] R.F. O'Doherty and N.A. Anstey, Geophys. Prospect. **19**, 430–458 (1971).

[234] R. Burridge, G. Papanicolaou and B. White, SIAM J. Appl. Math. **47**, 146–168 (1987).

[235] R. Burridge, G. Papanicolaou, P. Sheng and B. White, SIAM J. Appl. Math. **49**, 582–607 (1989).

[236] M. Asch, G. Papanicolaou, M. Postel, P. Sheng and B. White, Wave Motion **12**, 429–450 (1990).

[237] G. Papanicolaou, M. Postel, P. Sheng and B. White, Wave Motion **12**, 527–549 (1990).

[238] M.Asch, W. Kohler, G. Papanicolaou, M. Postel and B. White, SIAM Review **33**, 519–625 (1991).

[239] W. Kohler, G. Papanicolaou and B. White, Wave Motion **23**, 1–22 and 181–201 (1996).

[240] J.F. Clouet, J.P. Fouque and M. Postel, Wave Motion **22**, 145–170 (1995).

[241] J.F. Clouet and J.P. Fouque, Wave Motion **25**, 361 (1997).

[242] M. Fink, J. Phys. D: Appl. Phys. **26**, 1333–1350 (1993).

[243] S.R. Pal and A.I. Carswell, Appl. Opt. **12**, 1530 (1973).

[244] R.J. Allen, C.M.R. Platt, Appl. Opt. **16**, 3193 (1977).

[245] F. Nicolas, L.R. Bissonnette and P.H. Flamant, Appl. Opt. **36**, 3458 (1997).

[246] G. Zaccanti, P. Bruscaglioni and M. Dami: Appl. Opt. **29**, 3938 (1990).

[247] G.C. Mooradian, M. Geller, L.B. Stotts, D.H. Stephens and R. A. Krautwald: Appl. Opt. **18**, 429 (1979).

[248] F. Nicolas, Proc. 8th MUSCLE, Québec (1996).

[249] D.M. Winker, R.H. Couch and M.P. McCormick, Proc. IEEE **84**, No. 2, 164–180 (1996).

[250] D.M. Winker, Proc. 8th MUSCLE, Québec (1996), p. 1.

[251] A. Davis, A. Marshak, R. Cahalan and W. Wiscombe, J. Atmos. Sci., **54**, 241 (1997).

[252] L.R. Bissonnette, P. Bruscaglioni, A. Ismaelli, G. Zaccanti, A. Cohen, Y. Benayahu, R.D. Harack, L.D. Cohen, C. Flesia, P. Schwendimann, M. Oppel, D.M. Winkel, E.P. Zege, I.L. Katsev and I.N. Polonsky, Appl. Phys. B **60**, 355–362 (1995).

[253] P. Bruscaglioni, A. Ismaelli, G. Zaccanti, Appl. Phys. B, **60**, 325 (1995).

[254] A.V. Starkov, M. Noormohammadian and U. G. Oppel: Appl. Phys. B, **60**, 335 (1995).

[255] G. Zaccanti, P. Bruscaglioni, M. Gurioli and P. Sansoni, Appl. Opt. **32**, 1590 (1993).

[256] M. Gai, M. Gurioli, P. Bruscaglioni, A. Ismaelli and G. Zaccanti, Appl. Opt. **35**, 5435 (1996).

[257] P. Bruscaglioni and A. Ismaelli, Opt. Commun. **27**, 9 (1978).

[258] E.W. Eloranta, PhD dissertation, University of Wisconsin, Madison, Wisconsin (1972).

[259] E.W. Eloranta and S.T. Shipley "A solution for multiple scattering", in: *Atmospheric aerosols: their formation, optical properties and effects*, edited by A. Deepak (Spectrum Press, Hampton 1982).

[260] L. R. Bissonnette, Appl. Opt. **27**, 2478 (1988).

[261] L. R. Bissonnette, Appl. Opt. **35**, 6449 (1996).

[262] C. Flesia and P. Schwendimann, Appl. Phys. B **60**, 331 (1995).

[263] E.P. Zege, I.L. Katsev, I.N. Polonsky, Appl. Phys. B **60**, 345–353 (1995).

[264] I. L. Katsev, E.P. Zege, A. S. Prikhach and I.N. Polonsky, J. Opt. Soc. Am. A **14**, 6, 1338 (1997).

[265] E.P. Zege, A.P. Ivanov and I.L. Katsev, *Image Transfer through a Scattering Medium* (Springer, Berlin, Heidelberg ,1991).

[266] C. M. R. Platt, J. Atmos. Sci. **30**, 1191 (1973).

[267] L.R. Bissonnette and D.L. Hutt, Appl. Opt. **29**, 5045 (1990).

[268] G. Roy and L.R. Bissonnette, *Advances in Atmospheric Remote Sensing with Pidar: Selected papers of the 18th International Laser Radar Conference*, edited by A. Ansmann, R. Neuber, P. Rairoux and U. Wandinger (Springer, Berlin, Heidelberg, 1997).

[269] E.W. Eloranta, Proc. 8th MUSCLE, Québec (1996), p. 98.

[270] F. Nicolas, *Détermination des propriétés optiques et microphysiques des nuages par télédétection lidar en présence de diffusion multiple* Thesis Ecole Polytechnique, France (1997).

[271] A.G. Borovoi, Proc. 4th MUSCLE, Florence (1990).

[272] R. del Rio, S. Jitomirskaya, Y. Last and B. Simon, Phys. Rev. Lett. **75**, 117 (1995).

[273] R.E. Borland, Proc. Roy. Soc. Lond. A **274**, 529 (1963).

[274] N. Mott and W.D. Twose, Adv. Phys. **10**, 107 (1961).

[275] H. Kunz and B. Souillard, Commun. Math. Phys. **78**, 201 (1980).

[276] F. Delyon, Y.E. Lévy and B. Souillard, Phys. Rev. Lett. **55**, 618 (1985).

[277] D.J. Thouless, J. Phys. C **5**, 77 (1972).

[278] B. Simon, Commun. Math. Phys. **89**, 227 (1983).

[279] J.F. Fröhlich and T.M. Spencer, Commun. Math. Phys. **88**, 151 (1983).

[280] M. Aizenman and S. Molchanov, Commun. Math. Phys. **157**, 245 (1993).

[281] A. Figotin and A. Klein, J. Stat. Phys. **76**, 985 (1994).

[282] A. Figotin and A. Klein, J. Stat. Phys. **75**, 997–1201 (1994).

[283] J.M. Combes and P.D. Hislop, J. Funct. Anal. **124**, 149–180 (1994).

[284] K. Ishii and H. Matsuda, Suppl. Progr. Theor. Phys. **45**, 56–86 (1970).

[285] L.A. Pastur, Funct. Anal. Appl. **6**, 163–165 (1972).

[286] S. Kotani, in: *Proc. Taneguchi Int. Symp. on Stochastic Processes and Mathematical Physics*, edited by K. Ito. (North-Holland, Amsterdam, 1985), pp. 219–250.

[287] B. Simon and T. Wolff, Commun. Pure Appl. Math. **39**, 75–90 (1986).

[288] I.J. Goldsheid, S.A. Molcanov and L.A. Pastur, Funct. Anal. Appl. **11**, 1–10 (1977).

[289] J. Lacroix, Ann. Inst. H. Poincaré A **40**, 97–116 (1984).

[290] J. Fröhlich, F. Martinelli, E. Scoppola and T. Spencer, Commun. Math. Phys. **101**, 21–46 (1985).

[291] H. von Dreifus and A. Klein, Commun. Math. Phys. **124**, 285–299 (1989).

[292] R. Carmona, J. Funct. Anal. **51**, 229–258 (1983).

[293] Acoustic Waves: A. Figotin and A. Klein, Comm. Math. Phys. **180**(2), 439 (1996). Electromagnetic Waves: A. Figotin and A. Klein, Comm. Math. Phys. **184**(2), 411 (1997).

[294] F. Klopp, Commun. Math. Phys. **167**, 553–569 (1995).

[295] M. Reed and B. Simon, *Methods of Modern Mathematical Physics IV; Analysis of Operators* (Academic, New York, 1978).

[296] B. Simon, Commun. Math. Phys. **134**, 209–212 (1990).

[297] R. Carmona and J. Lacroix, *Spectral Theory of Random Schrödinger Operators*, (Birkhäuser, Boston, 1990).

[298] L. Pastur and A. Figotin, *Spectra of Random and Almost-Periodic Operators*, (Springer, Berlin, Heidelberg, 1991).

[299] H.J. Dorren and A. Tip, J. Math. Phys. **32**, 630 (1991).

[300] A. Tip, J. Math. Phys. **38**, 3545 (1997).

[301] R.G. Newton, *Scattering Theory of Waves and Particles* (McGraw-Hill, New York, 1966).

[302] K. Aki, Bull. Seism. Soc. Am. **82**, 1969–1972 (1992).

[303] V. Korneev and L. Johnson, preprint (1993).

[304] G.E. Backus, J. Geophys. Res. **67**, 4427–4440 (1962).

[305] R.F. O'Doherty and N.A Anstey, Geophys. Prospect. **19**, 430–458 (1971).

[306] N.C. Banik, I. Lerche and R.T. Shuey, Geophysics **50**, 2768–2774 (1985).

[307] M. Asch, W. Kohler, G. Papanicolaou, M. Postel and B. White, SIAM Review, Dec. 1991.

[308] B. White, P. Sheng and B. Nair, Geophysics **55**, 1158–1165 (1990).

[309] P. Sheng, B. White, Z.Q. Zhang and G. Papanicolaou, Phys. Rev. B **34**, 4757–4761 (1986).

[310] P. Lewicki, R. Burridge and G. Papanicolaou, Wave Motion **20**, 177–195 (1994).

[311] J. Chillan, PhD Thesis, Ecole Polytechnique, Paris, (1996).

[312] W. Kohler, G. Papanicolaou and B. White, Wave Motion **23**, 1 (1996); ibid. 181–201.

[313] A. Paul and F. Nicollin, Geophys. J. Int. **99**, 229–246 (1989).

[314] M. Campillo and A. Paul, J. Geophys. Res. **97**, 3405–3416 (1992).

[315] K. Aki and B. Chouet, J. Geophys. Res. **80**, 3322–3342 (1975).

[316] T.G. Rautian and I.R. Khalturin, Bull. Seism. Soc. Am. **68**, 923–948 (1978).

[317] P. Spudich P. and T. Bostwick, J. Geophys. Res. **92**, 10526–10546 (1987).

[318] F.D. Scherbaum, F. D. Gillard and N. Deichmann, Phys. Earth Planet. Inter. **67**, 137–161 (1991).

[319] M. Herraiz and A.M. Espinoza, Pure Appl. Geophys. **121**, 499–577 (1987)

[320] H. Sato, J. Phys. Earth **25**, 27–41 (1977).

[321] L.R. Jannaud, P.M. Adler and C.G. Jacquin, J. Geophys. Res. **96**, 18215–18231 (1991).

[322] H. Sato, J. Geophys. Res. **89**, 1221–1241 (1984).

[323] R.S. Wu, Geophys. J. Roy. Astro. Soc. **82**, 57–80 (1985).

[324] R.S. Wu and K. Aki, Pure Appl. Geophys. **128**, 49–80 (1988).

[325] K. Mayeda, F. Su and K. Aki, Phys. Earth Planet. Inter. **67**, 104–114 (1991).

[326] Y. Zeng, F. Su and K. Aki, J. Geoph. Res. **83**, 1264–1276 (1993); J. Geophys. Res. **96**, 607–619.

[327] A. Gusev and I.R. Abubakirov, Phys. Earth Planet. Inter. **49**, 30–36 (1987).

[328] M. Hoshiba, Phys. Earth Planet. Inter. **67**, 123–136 (1991).

[329] L. Margerin, M. Campillo and B.A. van Tiggelen, submitted to Geoph. J. Int. (1997).

[330] Y. Zeng, Bull. Seism. Soc. Am. **83**, 1264–1276 (1993).

[331] H. Sato, Geophys. J. Int. **117**, 487–494 (1994).

[332] R.L. Weaver, J. Mech. Phys. Solids **38**, 55–86 (1990).

[333] L.V. Ryjhik, G.C. Papanicolaou and J.B. Keller, Wave Motion **24**, 327 (1996).

[334] J. Turner, Bull. Seism. Soc. Am., in press (1997).

[335] C.A. Schultz and M.N. Toksoz, Geophys. J. Int. **114**, 91–102 (1993).

[336] C.A. Schultz and M.N. Toksoz, Geophys. J. Int. **117**, 783–810 (1994).

[337] M. Campillo, B. Feignier, M. Bouchon and N. Bethoux, J. Geophys. Res. **98**, 1987–1996 (1993).

[338] K. Aki and V. Ferrazzini, Geophys. Res. Lett. in press (1997).

[339] K. Kato, K. Aki and M. Takemura, Bull. Seism. Soc. Am. **85**, 467–477 (1995).

List of Contributors

Keiiti Aki
Observatoire Volcanologique de la Réunion, 14 Route nationale 3, 27ème km, F-97418 Pleine des Cafres, La Réunion, France

Kurt Busch
Department of Physics, University of Toronto, 60 St George Street, Toronto, Ontario, M5S 1A7, Canada

Michel Campillo
Laboratoire de Géophysique Interne et Tectonophysique, Université Joseph Fourier, BP 53X, F-38041 Grenoble, France

Arnaud Derode
Laboratoire Ondes et Acoustiques/CNRS, ESPCI, 10 rue Vauquelin, F-75231 Paris Cedex 05, France

Mathias Fink
Laboratoire Ondes et Acoustiques/CNRS, ESPCI, 10 rue Vauquelin, F-75231 Paris Cedex 05, France

Pierre Flamant
CNRS/Laboratoire de Météorologie Dynamique, Ecole Polytechnique, F-91128 Palaiseau, France

Jean-Pierre Fouque
CNRS/Centre de Mathématiques Appliquées, Ecole Polytechnique, F-91128 Palaiseau, France

Azriel Genack
Physics Department, Queens College, City University of New York, 65-30 Kissena Boulevard, Flushing, NY 11367-0904, USA

Jean Lacroix
Centre de Mathématiques Appliquées, Ecole Polytechnique, F-91128 Palaiseau, France

Ad Lagendijk
Van der Waals–Zeeman Laboratory, University of Amsterdam, Valckenierstraat 65–67, 1018 XE Amsterdam, The Netherlands

Guy Ledanois
Laboratoire de Physique des Milieux Ionisés/CNRS, Ecole Polytechnique,
F-91128 Palaiseau, Cedex, France

Georg Maret
Fakultät für Physik, University of Konstanz, POB 5560, D-78434 Konstanz,
Germany

Ludovic Margerin
Laboratoire de Géophysique Interne et Tectonophysique, Université Joseph
Fourier, BP 53X, F-38041 Grenoble, France

Roger Maynard
Laboratoire de Physique et Modélisation des Milieux Condensés/CNRS, Uni-
versité Joseph Fourier, Maison des Magistères, BP 166, F-38041 Grenoble
Cedex 9, France

François Nicolas
CNRS/Laboratoire de Météorologie Dynamique, Ecole Polytechnique, F-
91128 Palaiseau, France

Geert Rikken
Grenoble High Magnetic Field Laboratory, Max Planck Institut/CNRS, BP
166, F-38042 Grenoble, France

Philippe Roux
Laboratoire Ondes et Acoustiques/CNRS, ESPCI, 10 rue Vauquelin, F-75231
Paris Cedex 05, France

Patrick Sebbah
Laboratoire de Physique de la Matière Condensée, Université de Nice, Parc
Valrose, F-06108 Nice Cedex 02, France

Ping Sheng
Department of Physics, Hong Kong University of Science and Technology,
Clear Water Bay, Kowloon, Hong Kong

Costas Soukoulis
Ames Laboratory and Department of Physics and Astronomy, Iowa State
University, Ames, IA 50011, USA

Jean-Louis Thomas
Laboratoire Ondes et Acoustiques/CNRS, ESPCI, 10 rue Vauquelin, F-75231
Paris Cedex 05, France

Bart van Tiggelen
Laboratoire de Physique et Modélisation des Milieux Condensés/CNRS, Uni-
versité Joseph Fourier, Maison des Magistères, BP 166, F-38041 Grenoble
Cedex 9, France

Adriaan Tip
FOM-Instituut voor Atoom- en Molecuulfysica, Kruislaan 407, Amsterdam, The Netherlands

Jean Virmont
Laboratoire de Physique des Milieux Ionisés/CNRS, Ecole Polytechnique, F-91128 Palaiseau, Cedex, France

Diederik Wiersma
European Laboratory for Non-linear Spectroscopy, Large E Fermi, I-50125 Florence, Italy

Index

Springer Tracts in Modern Physics